101 Involved

ALGEBRA

Problems with Answers

$$\sqrt{x} + 5 = \sqrt{x + 225}$$

Chris McMullen, Ph.D.

101 Involved Algebra Problems with Answers
Chris McMullen, Ph.D.

www.improveyourmathfluency.com
www.monkeyphysicsblog.wordpress.com
www.chrismcmullen.com

Zishka Publishing
ISBN: 978-1-941691-92-2

Mathematics > Algebra

CONTENTS

INTRODUCTION

This book provides 101 involved algebra problems with answers and full solutions.

- The problems in this book are "<u>involved</u>." What does that mean? Some problems are involved in the sense that the solution requires several steps of algebra, while others are involved in terms of understanding, interpreting, analyzing, or applying ideas.
- The <u>level of difficulty</u> varies. A few problems should be relatively easy, a handful of problems should be quite challenging, and many problems fall somewhere in between. Since not all students have the same experience, knowledge, or ability, the problems will be easier for some students and more difficult for others.
- <u>Prerequisites</u>: Students should be familiar with standard algebra techniques, including how to solve a system of equations, the equation for a straight line, power rules (such as for $x^m x^n$), how to multiply expressions like $x\sqrt{x}(x^2 - \sqrt{x})$, and how to rationalize the denominator of an expression like $\frac{2}{\sqrt{x}}$ or like $\frac{3}{2-\sqrt{3}}$. The purpose of this book is to provide practice with a variety of algebra skills; this book is NOT intended to "teach" algebra (though it may be instructive to read the solution at the back of the book after attempting to solve each problem).
- Most students could benefit from additional algebra practice. The idea behind these problems is for students to practice applying a variety of algebra skills. The book also challenges students to solve problems out of context; it is easier to solve problems at the end of a chapter that is devoted to a particular topic than it is to solve a collection of problems that are not organized by topic.
- To check the answer to a problem without seeing the full solution, check the chapter entitled Just the Answers.
- Need a hint? Check the chapter entitled Hints.
- To find a solution to a problem, check the chapter entitled Full Solutions. One solution is presented for each problem. Keep in mind that there are often multiple ways to solve a given problem; the solution given may not be the only way to solve the problem. Some of the solutions also include explanations or comments.
- One problem is given on each page in order to provide adequate room to solve most of the problems. For a few problems, it may be desirable to have extra sheets of paper handy.
- An index of problems sorted by topic can be found at the back of the book.

Exercise [#]1

(A) Solve for x in the equation below.

$$1 + \sqrt{x} = \sqrt{4 + x}$$

(B) Verify each answer by plugging it into each side of the given equation.

Exercise #2

(A) Solve for x and y in the system of equations below.

$$\frac{1}{x} + \frac{1}{y} = \frac{1}{12}$$

$$x + y = 49$$

(B) Verify the answers by plugging them into the given equations.

Exercise #3

(A) Carry out the multiplication and division in the expression below, assuming that $x > 0$ and only allowing for positive roots. Simplify the answer as much as possible.

$$\frac{\left(x^{3/4} + x^{1/4}\right)\left(x^{2/3} - x^{1/6}\right)}{x^{5/12}}$$

(B) Plug $x = 4096$ into the final answer for Part A and show that this is consistent with the expression given in Part A.

Exercise $^{\#}4$

(A) There are three different straight lines that each pass through one pair of the following points: $(4, -4)$, $(-4, 8)$, and $(-8, -10)$. Write the equation of each of the three lines in the form $y = mx + b$, using numbers for m and b.

(B) Check the answers by plugging the (x, y) coordinates for each point into the two equations for the lines that pass through each point.

Exercise #5

(A) Solve for x in the equation below.

$$25 - \frac{9}{x^{3/2}} = \frac{11}{3}$$

(B) Explain how the last step in Part A can be solved without using a calculator. This question is asking specifically about the exponent of 3/2. What operation needs to be applied to each side of the equation in the last step, and how can the numerical value for this be determined without using a calculator?

(C) Verify the answer to Part A by plugging it into the given equation.

Exercise #6

(A) Solve for x, y, and z in the system of equations below.

$$14x + 21 - 36z - 33 + 22x = 0$$
$$54y - 21 + 36z + 5 - 24y = 0$$
$$x - y = z$$

(B) Verify the answers by plugging them into the given equations.

Exercise #7

(A) Multiply the expressions below and simplify the answer as much as possible.

$$(4x^2 - 12x + 9)\left(8x^2 - 6x + \frac{3x + 4}{2x - 3}\right)$$

(B) Plug $x = 2$ into the final answer for Part A and show that this is consistent with the expression given in Part A.

(C) Are there any values of x that may be a problem for the expression given in Part A? Explain.

Exercise #8

(A) Determine which values of the variables x and y satisfy the ratio $y{:}x = \sqrt{3}{:}2$ and also satisfy the equation below. Express each answer with a rational denominator.

$$x + y = \frac{1}{2}$$

(B) Verify the answers by plugging them into the given ratio and also plugging them into the given equation.

Exercise #9

(A) Apply algebra to show that the expressions for f and g below are equivalent for all possible values of the symbols x, y, z, p, q, and r.

$$f = (xq - yp)^2 + (xr - zp)^2 + (yr - zq)^2$$
$$g = (x^2 + y^2 + z^2)(p^2 + q^2 + r^2) - (xp + yq + zr)^2$$

(B) Plug $x = 1$, $y = 3$, $z = 5$, $p = 2$, $q = 4$, and $r = 6$ into the given equations for f and g to compare the expressions using numbers.

Exercise #10

(A) Solve for x and y in the system of equations below.

$$2.4x - 1.8y = 0.3$$
$$1.2x + 3.6y = 7.8$$

(B) Verify the answers by plugging them into the given equations.

Exercise #11

(A) Simplify the expression below, given that $w = 2x$.

$$u = x + \left(\frac{1}{w} + \frac{1}{x}\right)^{-1}$$

(B) Use the result from Part A to simplify the expression below.

$$t = x + \left(\frac{1}{u} + \frac{1}{x}\right)^{-1}$$

(C) Use the result from Part B to simplify the expression below.

$$s = x + \left(\frac{1}{t} + \frac{1}{x}\right)^{-1}$$

(D) Use the result from Part C to simplify the expression below.

$$r = x + \left(\frac{1}{s} + \frac{1}{x}\right)^{-1}$$

(E) Use the result from Part D to simplify the expression below.

$$q = x + \left(\frac{1}{r} + \frac{1}{x}\right)^{-1}$$

(F) Describe the pattern formed by the answers to Parts A-E.

Exercise #12

In the equations below, $x \neq 0$, $y \neq 0$, $y \neq x$, and $y \neq -x$.

$$x = \frac{t + u}{2}$$

$$y = \frac{t - u}{2}$$

Show that the equation $t^2 - u^2 = 4tz$ can be transformed into the equation below by making the change of variables above.

$$\frac{1}{x} + \frac{1}{y} = \frac{1}{z}$$

Exercise #13

(A) In the equation below, x, y, and z are three variables that could be measured in an experiment and k is a constant. The initial values of the variables are x_1, y_1, and z_1. The final values of the variables are x_2, y_2, and z_2.

$$xy = kz$$

Show that the formula below is valid if z is constant during the experiment.

$$x_1 y_1 = x_2 y_2$$

(B) Show that the formula below is valid if y is constant during the experiment.

$$\frac{x_1}{z_1} = \frac{x_2}{z_2}$$

(C) Show that the formula below is valid if x is constant during the experiment.

$$\frac{y_1}{z_1} = \frac{y_2}{z_2}$$

(D) Show that the formula below is valid in general, regardless of whether or not any variables are constant during the experiment.

$$\frac{x_1 y_1}{z_1} = \frac{x_2 y_2}{z_2}$$

Exercise #14

(A) Express the answer to the following subtraction as a single polynomial expression.
$$(3x^2 + 4x - 2)^3 - (2x^2 - 3x + 4)^3$$

(B) Plug $x = \frac{1}{2}$ into the final answer for Part A and show that this is consistent with the expression given in Part A.

Exercise #15

(A) In the equations below, $y \neq 0$.

$$z = \frac{x}{y}$$
$$w = \frac{xy}{2}$$

Derive an equation for w in terms of x and z only. The variable y may not appear in the final answer.

(B) Derive an equation for w in terms of y and z only. The variable x may not appear in the final answer.

(C) Given that $w < 9$ and $z = 2$, derive inequalities for the possible values of x and y.

Exercise #16

(A) Solve for w, x, y, and z in the system of equations below.

$$2w + 3x + 4y + 5z = 6$$
$$2x + 3y + 4z = 5$$
$$2y + 3z = 4$$
$$2z = 3$$

(B) Verify the answers by plugging them into the given equations.

(C) Comment on the answers to Part A.

Exercise #17

(A) Simplify the expression below as much as possible. Express the final answer with a rational denominator.

$$\frac{\sqrt{x}}{\sqrt{x} + \sqrt{3}} + \frac{\sqrt{3}}{\sqrt{x} - \sqrt{3}}$$

(B) For which value of x does the expression from Part A equal $\frac{5}{3}$?

(C) Check the answer to Part B by plugging the answer to Part B into the equation given in Part A.

(D) Are there any real values of x that may be a problem for the expression given in Part A? Explain.

Exercise #18

(A) Solve for x in the equation below, assuming that $\sqrt{x} > 0$.
$$16x - 3 = 8\sqrt{x}$$

(B) Verify the answer by plugging it into each side of the given equation.

Exercise #19

(A) In the equations below, $y \neq 0$.

$$x = \left(\frac{u + w}{2}\right) y$$

$$z = \frac{w - u}{y}$$

Derive an equation for w in terms of u, y, and z only. The variable x may not appear in the final answer.

(B) Derive an equation for x in terms of u, y, and z only. The variable w may not appear in the final answer.

(C) Derive an equation for w in terms of u, x, and z only. The variable y may not appear in the final answer.

Exercise #20

(A) Solve for x in the equation below.
$$(x + 7)(x - 3) = (x - 5)^2$$

(B) Verify the answer by plugging it into each side of the given equation.

Exercise #21

(A) Show that the expression below can be written in the form $\frac{P}{Q}$, where P and Q are each polynomials. Find P and Q.

$$\frac{2x}{x-3} - \frac{4-x}{x}$$

(B) Plug $x = 9$ into the final answer for Part A and show that this is consistent with the expression given in Part A.

(C) Are there any values of x that may be a problem for the expression given in Part A? Explain.

Exercise $^{\#}$22

(A) Solve for x and y in the system of equations below.

$$\frac{x - 6}{y + 3} = \frac{3}{4}$$

$$\frac{y + 7}{x + 9} = \frac{2}{3}$$

(B) Verify the answers by plugging them into each side of the given equations.

Exercise #23

(A) Use the equations and inequality below to derive an inequality for y in terms of x only. The variables u, w, and z may not appear in the final answer.

$$5y = 10x + \frac{z^2}{4}$$

$$w + 5 = \frac{u}{2}$$

$$u = \frac{z^2}{x}$$

$$w > 0$$

(B) Given that x is the radius of a circle, rewrite the answer to Part A in terms of the diameter of the circle. The answer to Part B should be an inequality for y in terms of the diameter only. The variables u, w, x, and z may not appear in the final answer.

(C) Derive an inequality for z in terms of y only. The variables u, w, x, and the diameter may not appear in the final answer. Assume that $z \geq 0$.

Exercise #24

(A) In the equation below, $x > 0$, $y > 0$, and $z > 0$. Show that the equation for z below can be written in the form $z = ax^b y^c$, where a, b, and c are constants. Determine the values of a, b, and c.

$$z = \left[\frac{(64x^6 y^9)^{2/3}}{(2x^2 y)^3 \sqrt{324 x^4 y^{10}}} \right]^{-\frac{5}{2}}$$

(B) Evaluate z for $x = 0.5$ and $y = 4$ in the final answer for Part A and show that this is consistent with the expression given in Part A.

Exercise #25

(A) Solve for x, y, and z in the system of equations below.

$$7x + 3y - z = 11$$
$$-4x + 9y - 3z = 22$$
$$3x - 5y + 2z = -21$$

(B) Verify the answers by plugging them into the given equations.

Exercise #26

(A) Solve for x in the equation below.

$$\frac{2}{25 - 4x} = \frac{x}{3}$$

(B) Verify each answer by plugging it into each side of the given equation.

Exercise #27

(A) The equations below represent three different straight lines. Determine the (x, y) coordinates of each of the three points where one pair of lines intersect.

$$y = x\sqrt{3} - \frac{7\sqrt{3}}{6}$$

$$y = -x\sqrt{3} - \frac{\sqrt{3}}{6}$$

$$y = -\frac{x\sqrt{3}}{3} + \frac{\sqrt{3}}{6}$$

(B) Check the answers by plugging the (x, y) coordinates for each point into the two equations for the lines that pass through each point.

Exercise #28

(A) Simplify the expression below until it has the form $\frac{P}{Q^k}$, where k is a constant and P and Q are polynomials. Determine the value of k and express P and Q terms of x.

$$\frac{1}{x + \sqrt{x^2 + 1}} + \left(\frac{1}{x + \sqrt{x^2 + 1}}\right)\left(\frac{x}{\sqrt{x^2 + 1}}\right)$$

(B) Plug $x = \sqrt{3}$ into the final answer for Part A and show that this is consistent with the expression given in Part A.

Exercise #29

(A) In the equations below, w, x, y, and z are each positive.

$$x + 2 = y + z$$
$$2x = 3y = wz$$

Use the equations above to derive the equations below.

$$x = 3(z - 2)$$
$$y = 2(z - 2)$$

(B) Use equations from Part A to derive the equation below.

$$z = \frac{12}{6 - w}$$

(C) Use equations and information from Parts A and B to derive an inequality for the possible values of w.

Exercise #30

(A) Show that the equation for y below can be written in the form $y = (ax + b)^2 + c$, where a, b, and c are constants. Determine the values of a, b, and c.

$$y = 36x^2 - 216x + 144$$

(B) Show that the equation for y below can be written in the form $y = (ax + b)^2 + c$, where a, b, c, p, q, and r are constants. Derive formulas for a, b, and c in terms of p, q, and r.

$$y = px^2 + qx + r$$

(C) Show that the formulas derived in Part B work for the values from Part A.

Exercise #31

(A) In the equations below, $p \neq 0$, $q \neq 0$, and $y \neq 0$.

$$z = \frac{2x}{y}$$

$$u = \frac{x}{p}$$

$$w = \frac{x}{q}$$

$$p + q = y$$

Derive an equation for z in terms of u and w only. The variables p, q, x, and y may not appear in the final answer.

(B) Suppose that $u = 20$, $w = 30$, and $x = 240$. Use these values to determine p, q, y, and z. Check the equations from Part A for consistency.

Exercise #32

(A) Solve for x in the equation below.

$$\left(\frac{1}{2}-\left(2-\left(1-\left(3-\left(\frac{1}{x}+\frac{1}{3}\right)^{-1}\right)^{-1}\right)^{-1}\right)^{-1}\right)^{-1}=\frac{3}{4}$$

(B) Verify the answer by plugging it into the given equation.

Exercise #33

(A) In the equation below, $a \neq 0$, $b \neq 0$, $c \neq 0$, and $r \neq 0$.

$$(x + a)(x + b)(x + c) = x^3 + px^2 + qx + r$$

Show that in order for the equation above to be true for all possible values of x, the values of a, b, and c must be related to the values of p, q, and r by the formulas below.

$$a + b + c = p$$
$$abc = r$$
$$\frac{1}{a} + \frac{1}{b} + \frac{1}{c} = \frac{q}{r}$$

(B) Plug the values $a = 2$, $b = 3$, $c = 4$, and $x = 5$ into the equations in Part A to check that the equations work for these numerical values.

Exercise #34

(A) Expand the expression below by multiplying expressions together. Rationalize the denominators in the answer. Express the final answer in the form $\left(P\sqrt{x} + Q\right)\frac{1}{x^2}$, where P and Q are each polynomials (without any radicals or fractional powers).

$$\left(\sqrt{x} - 1 + \frac{1}{\sqrt{x}}\right)^3$$

(B) Plug $x = 4$ into the final answer for Part A and show that this is consistent with the expression given in Part A.

(C) Are there any values of x that may be a problem for the expression given in Part A? Explain.

Exercise #35

(A) Solve for x, y, and z in the system of equations below.

$$x = \frac{9}{z}$$

$$y = \frac{16}{x}$$

$$z = \frac{25}{y}$$

(B) Verify the answers by plugging them into each side of the given equations.

Exercise #36

(A) Simplify the expression below until it has the form r^c. Express r and c in terms of k, p, and q only. The variables x and y may not appear in the final expressions for r and c.

$$\frac{y\left(\frac{q}{p}\right)^{\frac{1-k}{k}} - x}{y - x\left(\frac{p}{q}\right)^{\frac{1-k}{k}}}$$

(B) Are there any combinations of k, p, q, x, and y which may cause an exception in the simplification in Part A? If so, explain.

(C) Evaluate the answer to Part A for $k = \frac{3}{2}$, $p = 4$, and $q = 32$.

Exercise #37

(A) In the equations below, $w > 0$. Use the equations below to derive an equation for z in terms of c, k, and x, only. The variables w and y may not appear in the final answer.

$$ckx - z = \frac{ky^2}{w}$$

$$ckw(1 - x) = \frac{ky^2}{2}$$

(B) Are there are any values of x which guarantee that z will be zero, regardless of the values of c and k? If so, give the values of x that make z zero.

Exercise #38

(A) Write the equation of the straight line that passes through the points below in the form $y = mx + b$, using numbers for m and b.

$$\left(\frac{3\sqrt{3}}{4}, \frac{1}{2}\right)$$

$$\left(-\frac{\sqrt{3}}{4}, -\frac{1}{2}\right)$$

(B) Write the equation for the straight line that is perpendicular to the line in Part A and which passes through the midpoint of the points above.

(C) Plug the (x, y) coordinates for each of the three points into the equation of each line that passes through each point in order to check the answers for consistency.

Exercise #39

(A) In the equations below, $x > 0$, $y > 0$, and $z > 0$. Determine z when $y = 45$.

$$y = 50x - 5x^2$$
$$z = 50x\sqrt{3}$$

(B) Verify the answers to Part A by plugging the values into the given equations.

(C) In the system of equations given in Part A, determine the values of x, y, and z when the ratio of y to x equals 15. In Part C, y is no longer equal to 45, and the answer to Part A no longer applies. However, the two equations given in Part A do apply.

(D) Verify the answers to Part C by plugging the values into the equations given in Part A along with the ratio given in Part C.

Exercise #40

In the equations below, $p > 0$ and $y > 0$. Use the equations below to derive an equation for y in terms of x and c only. The symbols p, q, r, and z may not appear in the final answer.

$$pcx = \frac{1}{2}py^2 + \frac{1}{2}qz^2$$
$$q = \frac{2}{5}pr^2$$
$$y = rz$$

Exercise #41

(A) Solve for x and y in the system of equations below.

$$\frac{24}{x^2} = \frac{96}{y^2}$$

$$x + y = 18$$

(B) Verify the answers by plugging them into each side of the given equations.

Exercise #42

(A) Simplify the expression below as much as possible, assuming that x and y are each nonzero.

$$\frac{72x^5y^2 + 12x^4y^2 - 144x^3y^2}{15x^2y - 20xy}$$

(B) Plug $x = 2$ and $y = 3$ into the final answer for Part A and show that this is consistent with the expression given in Part A.

(C) Why would $x = 0$ or $y = 0$ be a problem? Are there any other values of x or y that may be a problem? If so, list the values that are problematic and explain why.

Exercise #43

(A) Solve for x in the equation below.

$$\left(x - \frac{1}{2}\right)^2 + x^2 + \left(x + \frac{3}{2}\right)^2 = \frac{15}{2}$$

(B) Verify the answers by plugging them into the given equation.

Exercise #44

(A) Solve for w, x, y, and z in the system of equations below.

$$wx^2y^3z^4 = 2^{30}$$
$$xy^2z^3 = 2^{14}$$
$$yz^2w^3 = 2^{26}$$
$$zw^2x^3 = 2^{30}$$

(B) Verify the answers by plugging them into the given equations.

Exercise #45

(A) Show that $(1 + x)(1 - x) = 1 - x^2$.

(B) Show that $(1 + x + x^2)(1 - x) = 1 - x^3$.

(C) Show that $(1 + x + x^2 + x^3)(1 - x) = 1 - x^4$.

(D) Show that $(1 + x + x^2 + x^3 + x^4)(1 - x) = 1 - x^5$.

(E) Using the pattern formed in Parts A-D, $1 + x + x^2 + x^3 + \cdots + x^n = \frac{P}{Q}$, where P and Q are polynomials. The three dots ($\cdots$) mean "and so on." The symbol n is the exponent of the final term. Express P and Q in terms of x and n.

(F) Are there any restrictions on the possible values of x in Part E? Explain.

Exercise #46

(A) In the equations below, $a > 0$, $b > 0$, and $c > 0$.

$$\sqrt{(x+c)^2 + y^2} + \sqrt{(x-c)^2 + y^2} = 2a$$
$$a^2 = b^2 + c^2$$

Use the equations above to derive the equation below.

$$\frac{x^2}{a^2} + \frac{y^2}{b^2} = 1$$

(B) Plug $a = 10$, $b = 6$, and $x = 5$ into the equation derived in Part A, and solve for y. Also solve for c.

(C) Plug $a = 10$, $b = 6$, $x = 5$, and the answers to Part B into the equations given in Part A to check the answers to Parts A and B for consistency.

Exercise #47

(A) Isolate ax in the inequality below, where a is a constant. The final answer should have ax isolated on one side of the inequality.

$$\frac{7}{4} - \frac{ax}{6} > \frac{5}{3}$$

(B) Plug $a = 2$ into the answer to Part A and isolate x.

(C) Plug $a = -2$ into the answer to Part A and isolate x.

(D) Comment on the final answers to Parts B and C.

(E) Use a calculator to check the answer to Part B as follows. Plug in a value that just barely satisfies the answer to Part B into the inequality given in Part A, and verify that it satisfies the original inequality. Plug in a value that just barely violates the answer to Part B into the inequality given in Part A, and verify that it does not work.

(F) Check the answer to Part C similarly to what was done in Part E.

Exercise $^\#$48

(A) Solve for x and y in the system of equations below.

$$5y + 27 = 12x - 8x^2$$
$$y + 9 = -4x$$

(B) Verify the answers by plugging them into each side of the given equations.

Exercise #49

In the equation below, $x > 0$ and $y > 0$. Show that the equation below is separable. This means to apply algebra to transform the equation into an equivalent equation where x only appears on one side of the equation and where y only appears on the other side of the equation. (The variables are not currently separated because x and y appear on both sides of the equation.)

$$16x^7y^2 + 9\sqrt{x^9y^4 - x^2y^4} = 25(x^6y^3 + x^6)^{2/3}$$

Exercise #50

(A) Variables $n, p, q, r, t, u, w, x, y$, and z are each positive.

$$u^2 = w^2 + y^2$$
$$x^2 = y^2 + z^2$$
$$q^2 + y^2 = t^2 + y^2 = n^2$$
$$w = p - q = r - t$$
$$z = r + t$$

Use the equations above to derive the equation below.

$$u^2 + x^2 = 2r^2 + 2n^2$$

(B) If $q = 5$, $n = 13$, and $r = 29$, determine p, t, u, w, x, y, and z.

(C) Verify that the given values and answers to Part B satisfy all of the equations in Part A.

Exercise #51

(A) Simplify the expression below as much as possible.

$$\left(\frac{x}{x+1} - \frac{x-1}{x}\right)^{-1} - \left(\frac{x}{x-1} - \frac{x+1}{x}\right)^{-1}$$

(B) Plug $x = 4$ into the final answer for Part A and show that this is consistent with the expression given in Part A.

(C) Are there any values of x that may be a problem for the expression given in Part A? Explain.

Exercise $^{\#}$52

In the equations below, $w \neq y$, $x \neq z$, and a and b are constants.

$$aw + bx = ay + bz$$
$$aw^2 + bx^2 = ay^2 + bz^2$$

Prove that if all of the conditions above are true, then the condition below must also be true.

$$w + y = x + z$$

Exercise #53

(A) Solve for x in the equation below.

$$\frac{6 - 2x}{3x + 9} = \frac{x - 10}{4x - 6}$$

(B) Verify each answer by plugging it into each side of the given equation.

Exercise #54

(A) In the equations below, $p > 0$, $q > 0$, $u > 0$, $w > 0$, and $z > 0$.

$$z = \sqrt{u^2 + (x - y)^2}$$

$$x = wp$$

$$y = \frac{1}{wq}$$

For fixed values of u, p, and q, what is the minimum possible value that z can have?

(B) Show that if z has the minimum possible value from Part A, the following condition must be true.

$$w = \frac{1}{\sqrt{pq}}$$

(C) Plug $u = 180$, $p = 18$, and $q = \frac{1}{450}$ into the answers to Parts A and B.

(D) Let $u = 180$, $p = 18$, and $q = \frac{1}{450}$. If $z = 225$, what is the value of w? (Note that the answers to Parts A, B, and C do NOT apply to Part D. However, the three equations given in Part A do apply to Part D.)

Exercise #55

In the equations below, $p > 0$, $x > 0$, $y > 0$, and $z > 0$.

$$z = \frac{ax}{y^2}$$

$$\frac{r^2}{z^2} = \frac{p^2}{x^2} + \frac{q^2}{y^2}$$

Use the equations above to derive the equation below.

$$r = \pm\frac{ap}{y^2}\sqrt{1 + \left(\frac{qx}{py}\right)^2}$$

Exercise #56

(A) Solve for x, y, and z in the system of equations below, assuming that $y \neq 0$.

$$x = yz$$
$$y = \left(z + \frac{3}{2}\right)x$$
$$z = (x + 3)y$$

(B) Verify the answers by plugging them into each side of the given equations.

Exercise #57

Show that the equation for z below can be written in the form $z = \frac{P}{Q^k}$, where P and Q are polynomials (without any radicals or fractional powers) and k is a constant. Determine the value of k and the expressions for P and Q.

$$z = \frac{4x^3\sqrt{x^2 + 4} - x^5(x^2 + 4)^{-1/2}}{x^2 + 4}$$

Exercise #58

(A) Solve for x, y, and z in the system of equations below.

$$x + \frac{1}{y} = 2$$

$$y + \frac{1}{z} = 1$$

$$z + \frac{1}{x} = 5$$

(B) Verify the answers by plugging them into each side of the given equations.

Exercise #59

In the equations below, $x > 0$, $y > 0$, and $z > 0$. Use the equations below to derive an equation for z in terms of a, c, k, and x only. The variables q, w, and y may not appear in the final answer.

$$k\frac{aq}{x^2} = qw$$

$$w = \frac{y^2}{x}$$

$$y = \frac{cx}{z}$$

Exercise #60

Use the equation below to show that z^2 can be written in the form $z^2 = x + w\sqrt{y}$, where w, x, and y are rational numbers. Determine the values of w, x, and y.

$$z = \sqrt{7 + \sqrt{5}} - \sqrt{7 - \sqrt{5}}$$

Exercise #61

(A) Solve for x in the equation below.

$$8 - \frac{4}{\sqrt{x}} = \frac{\sqrt{x}}{x}$$

(B) Verify the answer by plugging it into each side of the given equation.

Exercise #62

(A) Solve for w, x, y, and z in the system of equations below.

$$x = (4 + w)y$$
$$x = (4 - w)z$$
$$yz = 156$$
$$5x = 78w$$

(B) Verify the answers by plugging them into the given equations.

Exercise #63

(A) In the equations below, $|y + p|$ represents the absolute value of the sum of y and p.

$$\sqrt{x^2 + (y - p)^2} = |y + p|$$

$$a = \frac{1}{4p}$$

Use the equations above to derive the equation below.

$$y = ax^2$$

(B) Plug $a = \frac{1}{12}$ and $x = 18$ into the equation derived in Part A, and solve for y. Also solve for p.

(C) Plug $a = \frac{1}{12}$, $x = 18$, and the answers to Part B into the equations given in Part A to check the answers to Parts A and B for consistency.

Exercise #64

(A) In the second equation below, a, b, c, d, e, and f are constants.

$$y = \frac{5x + 60}{x^2 + 2x - 24}$$

Show that the equation above can be rewritten in the form below, and determine the values of a, b, c, d, e, and f.

$$y = \frac{a}{bx + c} + \frac{d}{ex + f}$$

(B) Are there any values of x that may be a problem for the equation given in Part A? Explain.

Exercise #65

(A) Solve for x and y in the system of equations below.

$$5x + 4y = 20$$
$$5x^2 + 4y^2 = 2180$$

(B) Verify the answers by plugging them into the given equations.

Exercise #66

In the equations below, $w \neq 0$, $z \neq 0$, and $k \neq 1$.

$$pw = qx$$
$$ry = sz$$
$$qx^k = ry^k$$
$$sz^k = pw^k$$

Prove that if all of the conditions above are true, then the condition below must also be true.

$$\frac{y}{z} = \frac{x}{w}$$

Exercise #67

(A) Isolate the variable in each inequality below.

$$\frac{1}{w} > \frac{5}{18} - \frac{7}{24} + \frac{1}{36} \quad , \quad \frac{1}{x} > \frac{3}{14} - \frac{8}{21} - \frac{1}{12}$$

$$\frac{1}{y} < \frac{8}{15} - \frac{3}{20} + \frac{7}{60} \quad , \quad \frac{1}{z} < \frac{9}{16} - \frac{5}{24} - \frac{23}{48}$$

(B) Comment on the final answers to Part A.

(C) Use a calculator to check each answer to Part A as follows. Plug in a value that just barely satisfies each answer to Part A into each inequality given in Part A, and verify that it satisfies the original inequality. Plug in a value that just barely violates each answer to Part A into the inequality given in Part A, and verify that it does not work.

Exercise #68

(A) In the equations below, $t, u, w, x, y,$ and z are each positive.

$$u^2 = w^2 + z^2$$
$$t^2 = w^2 + y^2$$
$$x + y = z$$

Use the equations above to derive the inequality below.

$$u^2 > t^2 + x^2$$

(B) Explain why the result of Part A has a greater than sign.

Exercise #69

(A) Simplify the expression below as much as possible, assuming that $y \neq 0$.

$$\frac{x}{y}(2 - z^2) + 2\left(z^2 - \frac{x}{y}\right) + z^2\left(\frac{x}{y} - 2\right)$$

(B) Plug $x = 8, y = 2$, and $z = 3$ into the expression given in Part A to check the answer to Part A.

(C) If $x = t^2 - u^2, y = \sqrt{p + q + r}$, and $z = 5v^{3/2}w^{1/2}$ are plugged into the expression given in Part A, will the answer to Part A be the same or different? Explain.

Exercice $^\#$70

(A) Solve for x in the equation below.
$$(x - 5)^3 = (x - 10)^2(x + 25)$$

(B) Verify each answer by plugging it into each side of the given equation.

Exercise #71

(A) The equations below represent one parabola and one straight line. Determine the (x, y) coordinates of each of the two points where the parabola and line intersect.

$$y = 2(x - 3)^2 - 5$$

$$y = 4x - \frac{13}{2}$$

(B) Check the answers by plugging the (x, y) coordinates for each point into each of the given equations.

(C) Sketch the parabola and line. Label the coordinates of the points of intersection on the sketch.

Exercise #72

(A) In the equations below, $c > 0$, $k > 0$, $u > 0$, and $y > 0$.

$$w = kyu$$

$$y = \frac{1}{\sqrt{1 - \dfrac{u^2}{c^2}}}$$

Use the equations above to derive the equation below. Challenge: Strive to logically combine equations above to arrive at the equation below (rather than merely plugging equations from above into the equation below to show that it works).

$$w^2 c^2 + k^2 c^4 = y^2 k^2 c^4$$

(B) Plug an equation from Part A into the equation below to show that the equation below is true. (Unlike Part A's challenge, it is not necessary to derive this equation.)

$$\frac{y^2}{y^2 - 1} = \frac{c^2}{u^2}$$

(C) Assuming that y is real and finite, give an inequality for the allowed values of u in terms of c. Also give an inequality for the possible values for y.

Exercise #73

(A) Solve for x in the equation below.

$$\sqrt{1 + \sqrt{4 + \sqrt{21 + \sqrt{x + \sqrt{3x^2 - 55x + 267}}}}} = 2$$

(B) Verify each answer by plugging it into the given equation.

Exercise #74

(A) Solve for x and y in the system of equations below.

$$\frac{2x}{3} - \frac{3y}{4} = \frac{1}{6}$$
$$\frac{5x}{7} + \frac{9y}{4} = \frac{17}{4}$$

(B) Verify the answers by plugging them into the given equations.

Exercise #75

(A) In the equations below, $c, k, p,$ and w are each positive.

$$x^2 + y^2 = 1$$

$$px = ck$$

$$py = \frac{ku^2}{w}$$

Derive an equation for p in terms of $c, k, u,$ and w only. The variables x and y may not appear in the final answer.

(B) Derive an equation for w in terms of $c, u,$ and x only. The symbols $k, p,$ and y may not appear in the final answer.

Exercise #76

(A) The equations below represent one ellipse and one hyperbola. Determine the (x, y) coordinates of each of the **four** points where the ellipse and hyperbola intersect.

$$\frac{x^2}{9} + \frac{y^2}{4} = 1$$

$$y = \frac{\sqrt{5}}{x}$$

(B) Check the answers by plugging the (x, y) coordinates for each point into each of the given equations.

(C) Sketch the ellipse and hyperbola. Label the coordinates of the points of intersection on the sketch.

Exercise #77

(A) Solve for x in the equation below.

$$\left(3 - \sqrt{x}\right)^2 - 15 = 2\sqrt{x} - 6$$

(B) Verify each answer by plugging it into each side of the given equation.

Exercise #78

(A) The two equations below have three variables: x, y, and z. Ordinarily, given two equations with three variables, it would not be possible to solve for any of the variables. However, for the pair of equations below, it is actually possible to solve for one of the three variables. Which of the three variables can be determined? What is special about the pair of equations below that makes it possible to solve for one variable?

$$6x + 2y = 5 + 9z$$
$$12z - 6y = 8x - 11$$

(B) Solve the system of equations above for the one variable that can be determined.

(C) For the two variables that are indeterminate, derive a simple equation that relates the two variables. The only variables in this equation should be the two variables that are indeterminate.

Exercise #79

(A) In the equations below, $a > 0$, $b > 0$, and $c > 0$. Determine the values of a, b, and c needed in order for the two equations for y below to be equivalent.

$$y = a^2 - (b - cx)^2$$
$$y = 30x - 5x^2$$

(B) Use the results from Part A to determine the maximum possible value for y.

(C) Which values of x make y equal the value in Part B?

Exercise #80

(A) Solve for x, y, and z in the system of equations below.

$$6z - 4y = 3$$

$$3y = x - \frac{1}{4}$$

$$\frac{z}{6} = \left(x - \frac{2}{3}\right)^2$$

(B) Verify the answers by plugging them into each side of the given equations.

Exercise #81

(A) Show that the expression below can be written in the form $\frac{P}{Q}$, where P and Q are each polynomials. Find P and Q.

$$\frac{3x + 2}{2x - 1} - \frac{4x + 5}{6x + 8} + \frac{2}{x}$$

(B) Plug $x = 2$ into the final answer for Part A and show that this is consistent with the expression given in Part A.

(C) Are there any values of x that may be a problem for the expression given in Part A? Explain.

Exercise #82

(A) Solve for x and y in the system of equations below.

$$\sqrt{(x-3)^2 + (y-8)^2} = 5$$
$$3y = 4x + 12$$

(B) Verify the answers by plugging them into the given equations.

(C) Offer a geometric interpretation for the answers to Part A.

Exercise #83

(A) In the equations below, $a \neq 0$, $b \neq 0$, $c \neq 0$, and $b \neq -a$.

$$w = ax^2 + by^2$$

$$\frac{1}{a} + \frac{1}{b} = \frac{1}{c}$$

$$ax + by = 0$$

$$z = x - y$$

Use the equations above to derive the equations below.

$$x = \frac{bz}{a + b}$$

$$y = -\frac{az}{a + b}$$

(B) Use equations from Part A to derive the equation below.

$$w = cz^2$$

Exercise $^{\#}84$

(A) Solve for x and y in the system of equations below, assuming that $\sqrt{x} > 0$ and $\sqrt{y} > 0$.

$$3\sqrt{xy} = 28$$
$$8\sqrt{x} - 9\sqrt{y} = 4$$

(B) Verify the answers by plugging them into the given equations.

Exercise #85

(A) In the equations below, $x \neq 0$, $y \neq 0$, and $z \neq 0$.

$$xy^2 + yz^2 + zx^2 = a$$
$$xz^2 + yx^2 + zy^2 = b$$
$$xyz = c$$

Use the equations above to derive the equation below.

$$(x + y + z)\left(\frac{1}{x} + \frac{1}{y} + \frac{1}{z}\right) = \frac{a + b + 3c}{xyz}$$

(B) Plug $x = 2$, $y = 3$, and $z = 6$ into the equations given in Part A. Determine a, b, and c for these values of x, y, and z.

(C) Plug $x = 2$, $y = 3$, and $z = 6$ and the answers for a, b, and c from Part B into the result derived in Part A to check for consistency.

Exercise #86

(A) Although it is not possible to solve for x or y in the equation below, it is possible to determine the ratio of x to y for this equation. Determine the ratio of x to y for the equation below.

$$4x^2 + 6xy = 5y^2$$

(B) Verify each answer by plugging it into each side of the given equation.

Exercise #87

(A) The formulas for the volume and surface area of a sphere are given below in terms of the radius of the sphere, where π is a constant.

$$V_{\text{sphere}} = \frac{4}{3}\pi R^3 \quad , \quad S_{\text{sphere}} = 4\pi R^2$$

The formulas for the volume and surface area of a cube are given below in terms of the edge length of the cube.

$$V_{\text{cube}} = L^3 \quad , \quad S_{\text{cube}} = 6L^2$$

For a cube and sphere that have the same volume, derive an equation for the radius of the sphere in terms of the edge length of the cube. The final answer should have R isolated on one side, L on the other side, the constant π, and numbers. The symbols for volume and surface area should not appear in the final answer. Do not replace π with a numerical value.

(B) For a cube and sphere that have the same volume, derive an equation for the ratio of the surface area of the sphere to the surface area of the cube. The final answer should have $\frac{S_{\text{sphere}}}{S_{\text{cube}}}$ isolated on one side, the constant π, and numbers. The symbols for radius, edge length, and volume should not appear in the final answer. Do not replace π with a numerical value.

(C) Without using a calculator, use the answer to Part B to determine whether the cube or sphere (of the same volume) has less surface area.

Exercise #88

(A) Use the proportions $x{:}z = w{:}x$ and $y{:}z = (z - w){:}y$ to derive the equation below.
$$x^2 + y^2 = z^2$$

(B) Plug $z = 25$ and $w = 9$ into the proportions from Part A, and use the proportions from Part A to determine x and y.

(C) Plug the values from Part B into the equation derived in Part A, and verify that it agrees with $z = 25$.

Exercise #89

(A) Solve for x in the equation below, assuming that x is real.

$$\left(256 + \left(2 + \left(7 + \left(3 + (2x^2 - 27x + 44)^{\frac{1}{3}}\right)^2\right)^{\frac{1}{5}}\right)^4\right)^{\frac{1}{3}} = 8$$

(B) Verify each answer by plugging it into each side of the given equation.

Exercise #90

(A) In the equations below, $a > 0$, $b > 0$, and $y > t > u > x > 0$.

$$\frac{t}{b(y-t)} + \frac{u}{a(u-x)} - \frac{x(t-u)}{a(u-x)^2} = 0$$

$$\frac{t}{b(y-t)} + \frac{u}{a(u-x)} - \frac{y(t-u)}{b(y-t)^2} = 0$$

Prove that if all of the conditions above are true, then the condition below must also be true.

$$\frac{u}{t} = \sqrt{\frac{x}{y}}$$

(B) Use the equations from Part A to derive the equations below.

$$t = \frac{\sqrt{axy} + y\sqrt{b}}{\sqrt{a} + \sqrt{b}}$$

$$u = \frac{x\sqrt{a} + \sqrt{bxy}}{\sqrt{a} + \sqrt{b}}$$

Exercise #91

This problem uses the equations from Exercise 90.

$$w = \frac{t - u}{\dfrac{t}{b(y - t)} + \dfrac{u}{a(u - x)}}$$

Use equations from Exercise 90 and the equation above to derive the equations below.

$$\frac{t}{b(y - t)} + \frac{u}{a(u - x)} = \frac{\sqrt{xy}(t - u)}{\sqrt{ab}(u - x)(y - t)}$$

$$u - x = \frac{\sqrt{x}(\sqrt{ax} + \sqrt{by}) - x(\sqrt{a} + \sqrt{b})}{\sqrt{a} + \sqrt{b}}$$

$$y - t = \frac{y(\sqrt{a} + \sqrt{b}) - \sqrt{y}(\sqrt{ax} + \sqrt{by})}{\sqrt{a} + \sqrt{b}}$$

$$w = ab \left(\frac{\sqrt{y} - \sqrt{x}}{\sqrt{a} + \sqrt{b}} \right)^2$$

Exercise #92

(A) In the equations below, $a \neq 0$ and $y \neq 0$.

$$ax^2 + ay + bx + c = 0$$
$$2ax\sqrt{y} + b\sqrt{y} = 0$$

Derive an equation for x in terms of a and b only. The symbols c and y may not appear in the final answer.

(B) Derive an equation for y in terms of a, b, and c only. The variable x may not appear in the final answer.

(C) How do the parts of this problem relate to the standard quadratic formula?

Exercise #93

(A) Isolate the variable in each inequality below.

$$\frac{3}{w} > \frac{1}{w-2} \quad , \quad \frac{8}{x} > \frac{5}{x+3}$$

$$\frac{4}{y} < \frac{3}{y-6} \quad , \quad \frac{9}{z} < \frac{7}{z+4}$$

(B) Comment on the final answers to Part A.

(C) Use a calculator to check each answer to Part A as follows. Plug in a value that just barely satisfies each answer to Part A into each inequality given in Part A, and verify that it satisfies the original inequality. Plug in a value that just barely violates each answer to Part A into the inequality given in Part A, and verify that it does not work.

Compare with the book's answers. There are more solutions than most students expect.

Exercise #94

(A) Find the equation of the line that is tangent to the circle $x^2 + y^2 = 25$ for which $x = 3$ and $y > 0$. Express the answer in the form $y = mx + b$, using numbers for m and b. The tangent line intersects the circle only at a single point.

(B) Plug the (x, y) coordinates for the point where the line is tangent to the circle into the equation of the line and the equation of the circle in order to check for consistency.

(C) Verify that the equation of the tangent line is perpendicular to the equation of the radius that connects to the point where the line is tangent to the circle.

Exercise #95

(A) In the equations below, $|x - y|$ represents the absolute value of $x - y$. Find every combination of x and y that satisfies the pair of equations below.

$$\frac{|x - y|}{x} = 0.1$$

$$3x + 6y = 8.4$$

(B) Verify each set of answers by plugging them into the given equations.

Exercise #96

(A) In the equation below, a, b, c, and d are real constants, and $a > 0$.

$$y = ax^3 + bx^2 + cx + d$$

Plug $x = t - \dfrac{b}{3a}$ into the equation above. Show that this transforms the equation above into the equation below. Derive equations relating p and q to a, b, c, and d.

$$y = at^3 + pt + q$$

(B) Plug $y = \dfrac{z}{\sqrt{a}} + q$ and $t = \dfrac{u}{\sqrt{a}}$ into $y = at^3 + pt + q$. Show that this transforms $y = at^3 + pt + q$ into the equation below.

$$z = u^3 + pu$$

Exercise #97

(A) In the equation below, $z > 0$. Show that the equation for z below can be written in the form $z = \sqrt{x} - \sqrt{y}$, where x and y are rational numbers (but $\sqrt{x}$ and $\sqrt{y}$ are not). Determine the values of x and y.

$$z = \sqrt{\frac{1}{2} - \frac{\sqrt{3}}{4}}$$

(B) Express z in standard form with a rational denominator and without any nested roots.

(C) Verify that the answer to Part B is consistent with the equation given in Part A.

Exercise #98

In the equations below, $c > 0$, $k > 0$, $x \neq 0$, and $y \neq 0$.

$$x = yq + zs$$
$$0 = yr - zt$$
$$x + kc = y + \sqrt{z^2 + k^2c^2}$$
$$q^2 + r^2 = s^2 + t^2 = 1$$

Use the equations above to derive the equation below. Challenge: Strive to logically combine equations above to arrive at the equation below (rather than merely plugging equations from above into the equation below to show that it works).

$$\frac{1}{y} - \frac{1}{x} = \frac{1 - q}{kc}$$

Exercise #99

(A) Solve for w, x, y, and z in the system of equations below.

$$\frac{1}{x} + \frac{1}{y} = \frac{1}{3}$$

$$\frac{1}{w} + \frac{1}{z} = \frac{1}{4}$$

$$w + y = 18$$

$$z - x = 8$$

(B) Verify the answers by plugging them into the given equations.

Exercise #100

(A) Variables t, u, w, x, y, and z are related by the equations below.

$$2wy = u^2 + w^2 - x^2$$
$$t^2 = u^2 - y^2$$
$$z = \frac{wt}{2}$$

Combine the equations above to derive the equation below.

$$z = \sqrt{\frac{4u^2w^2 - (u^2 + w^2 - x^2)^2}{16}}$$

(B) Show that the answer to Part A can be rewritten as:

$$z = \sqrt{\frac{(2uw + u^2 + w^2 - x^2)(2uw - u^2 - w^2 + x^2)}{16}}$$

(C) Show that the answer to Part B can be rewritten as:

$$z = \sqrt{\frac{[(u + w)^2 - x^2][x^2 - (u - w)^2]}{16}}$$

(D) Show that the answer to Part C can be rewritten as:

$$z = \sqrt{\frac{(u + w + x)(u + w - x)(u + x - w)(w + x - u)}{16}}$$

(E) Let $2s = u + w + x$. Show that the answer to Part D can be rewritten as:

$$z = \sqrt{s(s - u)(s - w)(s - x)}$$

Exercise #101

(A) In the equation below, t, u, x, and y are real numbers, i is the imaginary number, and z is a complex number. It is not necessary to know about complex numbers to solve this problem, other than to use the relation $i^2 = -1$ as part of the solution.

$$z = (x + iy)(t + iu)$$

Show that the equation above can be transformed into the equation below, and derive equations relating p and q to x and y.

$$z = (t + u)x(1 + i) - pu + iqt$$

(B) Given that a computer can add or subtract two numbers faster than it can multiply two numbers, explain how this transformation can help coders write programs that can carry out faster complex calculations.

Suggestion

Place a folded sheet of paper here to cover up the answers to problems that you have not yet attempted. This way, you will be less likely to accidentally see the answer to a problem that you are not yet ready to see.

Answer to Exercise #1

$$x = \frac{9}{4} = 2.25$$

Answers to Exercise #2

$$(1)\ x = 21 \text{ and } y = 28 \quad , \quad (2)\ x = 28 \text{ and } y = 21$$

Answers to Exercise #3

$$(A)\ x - 1 \quad (B)\ 4095$$

Answers to Exercise #4

$$y = -\frac{3}{2}x + 2 \quad , \quad y = \frac{x}{2} - 6 \quad , \quad y = \frac{9}{2}x + 26$$

Answer to Exercise #5

$$x = \frac{9}{16} = 0.5625$$

Answers to Exercise #6

$$x = \frac{1}{2} \quad , \quad y = \frac{1}{3} \quad , \quad z = \frac{1}{6}$$

Answers to Exercise #7

$$(A)\ 32x^4 - 120x^3 + 144x^2 - 51x + 4x \quad (B)\ 30 \quad (C)\ x = \frac{3}{2}$$

Answers to Exercise #8

$$x = 2 - \sqrt{3} \left(\text{equivalent to } \frac{1}{2 + \sqrt{3}} \right) \quad , \quad y = \sqrt{3} - \frac{3}{2} = \frac{2\sqrt{3} - 3}{2} = (2 - \sqrt{3})\frac{\sqrt{3}}{2}$$

Answers to Exercise #9

(A) $x^2q^2 + x^2r^2 + y^2p^2 + y^2r^2 + z^2p^2 + z^2q^2 - 2xypq - 2xzpr - 2yzqr$ (B) 24

Answers to Exercise #10

$$x = \frac{7}{5} = 1.4 \quad , \quad y = \frac{17}{10} = 1.7$$

Answers to Exercise #11

(A) $u = \frac{5x}{3}$ (B) $t = \frac{13x}{8}$ (C) $s = \frac{34x}{21}$ (D) $r = \frac{89x}{55}$ (E) $q = \frac{233x}{144}$

Answer to Exercise #12

$$\frac{1}{x} + \frac{1}{y} = \frac{1}{z}$$

Answers to Exercise #13

(A) $x_1y_1 = x_2y_2$ (B) $\frac{x_1}{z_1} = \frac{x_2}{z_2}$ (C) $\frac{y_1}{z_1} = \frac{y_2}{z_2}$ (D) $\frac{x_1y_1}{z_1} = \frac{x_2y_2}{z_2}$

Answers to Exercise #14

(A) $19x^6 + 144x^5 - 12x^4 + 91x^3 - 264x^2 + 192x - 72$ (B) $\frac{27}{64} - 27 = -\frac{1701}{64}$

Answers to Exercise #15

$$(A)\ w = \frac{x^2}{2z} \quad (B)\ w = \frac{y^2z}{2}$$

(C) $-6 < x < 6$, $-3 < y < 3$ (equivalent to $6 > x > -6$, $3 > y > -3$)

Answers to Exercise #16

$$w = -\frac{1}{16} = -0.0625 \ , \quad x = -\frac{1}{8} = -0.125 \ , \quad y = -\frac{1}{4} = -0.25 \ , \quad z = \frac{3}{2} = 1.5$$

Answers to Exercise #17

$$(A)\ \frac{x+3}{x-3} \quad (B)\ x = 12 \quad (D)\ x = 3$$

Answer to Exercise #18

$$x = \frac{9}{16} = 0.5625 \quad \left(x = \frac{1}{16} \text{ would require } \sqrt{x} = -\frac{1}{4} < 0\right)$$

Answers to Exercise #19

(A) $w = yz + u$ (B) $x = \frac{y^2 z}{2} + uy = \left(\frac{yz}{2} + u\right) y = \left(\frac{yz + 2u}{2}\right) y$ (C) $w = \pm\sqrt{u^2 + 2xz}$

Answer to Exercise #20

$$x = \frac{23}{7}$$

Answers to Exercise #21

(A) $P = 3x^2 - 7x + 12$, $Q = x^2 - 3x$ (B) $\frac{32}{9}$ (C) $x = 0$, $x = 3$

Answers to Exercise #22

$$x = 15 \quad , \quad y = 9$$

Answers to Exercise #23

(A) $y > \frac{5x}{2}$ or $y > 2.5x$ $\left(\text{equivalent to } \frac{5x}{2} < y \text{ or } 2.5x < y\right)$

(B) $y > \frac{5D}{4}$ or $y > 1.25D$ $\left(\text{equivalent to } \frac{5D}{4} < y \text{ or } 1.25D < y\right)$

(C) $2\sqrt{y} < z$ $\left(\text{equivalent to } 4y < z^2, z > 2\sqrt{y}, \text{ or } z^2 > 4y\right)$

Answers to Exercise #24

(A) $a = 243$, $b = 10$, $c = 5$ (B) 243

Answers to Exercise #25

$$x = \frac{11}{25} = 0.44 \quad , \quad y = -\frac{162}{25} = -6.48 \quad , \quad z = -\frac{684}{25} = -27.36$$

Answers to Exercise #26

$$(1) \ x = \frac{1}{4} = 0.25 \quad , \quad (2) \ x = 6$$

Answers to Exercise #27

$$\left(\frac{1}{2}, -\frac{2\sqrt{3}}{3}\right) \quad , \quad \left(1, -\frac{\sqrt{3}}{6}\right) \quad , \quad \left(-\frac{1}{2}, \frac{\sqrt{3}}{3}\right)$$

$$\text{Notes: } \frac{2\sqrt{3}}{3} = \frac{2}{\sqrt{3}} \quad , \quad \frac{\sqrt{3}}{6} = \frac{1}{2\sqrt{3}} \quad , \quad \frac{\sqrt{3}}{3} = \frac{1}{\sqrt{3}}$$

Answers to Exercise #28

$$\text{(A) } P = 1 \quad , \quad Q = x^2 + 1 \quad , \quad k = \frac{1}{2} \quad \text{(B) } \frac{1}{2}$$

Answers to Exercise #29

$$\text{(A) } x = 3(z - 2) \quad , \quad y = 2(z - 2) \quad \text{(B) } z = \frac{12}{6 - w}$$

$$\text{(C) } 6 > w > 0 \quad \text{(equivalent to } 0 < w < 6)$$

Answers to Exercise #30

$$\text{(A) } a = 6 \quad , \quad b = -18 \quad , \quad c = -180$$

$$\text{(B) } a = \sqrt{p} \quad , \quad b = \frac{q\sqrt{p}}{2p} = \frac{q}{2\sqrt{p}} \quad , \quad c = r - \frac{q^2}{4p}$$

Answers to Exercise #31

$$\text{(A) } z = \frac{2uw}{u + w} \left(\text{equivalent to } \frac{1}{u} + \frac{1}{w} = \frac{2}{z}\right)$$

$$\text{(B) } p = 12 \quad , \quad q = 8 \quad , \quad y = 20 \quad , \quad z = 24$$

Answer to Exercise #32

$$x = \frac{51}{16} = 3.1875$$

Answers to Exercise #33

$$\text{(A) } a + b + c = p \quad , \quad abc = r \quad , \quad \frac{1}{a} + \frac{1}{b} + \frac{1}{c} = \frac{q}{r} \quad \text{(B) } 504$$

Answers to Exercise #34

$$(A)\ P = x^3 + 6x^2 + 6x + 1 \quad, \quad Q = -3x^3 - 7x^2 - 3x \quad (B)\ \frac{27}{8} \quad (C)\ x = 0$$

Answers to Exercise #35

$$(1)\ x = \frac{12}{5} \quad, \quad y = \frac{20}{3} \quad, \quad z = \frac{15}{4} \quad (2)\ x = -\frac{12}{5} \quad, \quad y = -\frac{20}{3} \quad, \quad z = -\frac{15}{4}$$

Answers to Exercise #36

$$(A)\ r = \frac{q}{p} \quad, \quad c = \frac{1-k}{k}\ \left(\text{equivalent to } c = \frac{1}{k} - 1\right)$$

$$(B)\ k = 0 \quad, \quad p = 0 \quad, \quad q = 0 \quad, \quad y = x\left(\frac{p}{q}\right)^{\frac{1-k}{k}} \quad (C)\ \frac{1}{2}$$

Answers to Exercise #37

$$(A)\ z = ck(3x - 2) \quad (B)\ x = \frac{2}{3}$$

Answers to Exercise #38

$$(A)\ y = \frac{\sqrt{3}}{3}x - \frac{1}{4}\ \left(\text{equivalent to } y = \frac{x}{\sqrt{3}} - \frac{1}{4}\right) \quad (B)\ y_\perp = -x\sqrt{3} + \frac{3}{4}$$

Answers to Exercise #39

$$(A)\ (1)\ z = 50\sqrt{3} \quad (2)\ z = 450\sqrt{3} \quad (C)\ x = 7 \quad, \quad y = 105 \quad, \quad z = 350\sqrt{3}$$

Answer to Exercise #40

$$y = \sqrt{\frac{10cx}{7}}\ \left(\text{equivalent to } y = \frac{\sqrt{70cx}}{7}\right)$$

Answers to Exercise #41

$$(1)\ x = 6 \quad, \quad y = 12 \quad (2)\ x = -18 \quad, \quad y = 36$$

Answers to Exercise $^\#42$

$$\text{(A)} \;\; \frac{12x^2y(2x+3)}{5} = \frac{24x^3y + 36x^2y}{5} \qquad \text{(B)} \;\; \frac{1008}{5} = 201.6$$

$$\text{(C)} \; x = 0 \;\;, \;\; y = 0 \;\;, \;\; x = \frac{4}{3}$$

Answers to Exercise $^\#43$

$$\text{(1)} \; x = -\frac{5}{3} \quad \text{(2)} \; x = 1$$

Answers to Exercise $^\#44$

$$\text{(1)} \; w = 128, x = 32, y = 8, \text{ and } z = 2 \quad \text{(2)} \; w = -128, x = 32, y = -8, \text{ and } z = 2$$

$$\text{(3)} \; w = 128, x = -32, y = 8, \text{ and } z = -2$$

$$\text{(4)} \; w = -128, x = -32, y = -8, \text{ and } z = -2$$

Answers to Exercise $^\#45$

$$\text{(A)} \; 1 - x^2 \quad \text{(B)} \; 1 - x^3 \quad \text{(C)} \; 1 - x^4 \quad \text{(D)} \; 1 - x^5$$

$$\text{(E)} \; P = 1 - x^{n+1} \;\;, \;\; Q = 1 - x \quad \text{(F)} \; x = 1$$

Answers to Exercise $^\#46$

$$\text{(A)} \; \frac{x^2}{a^2} + \frac{y^2}{b^2} = 1 \quad \text{(B)} \; y = \pm 3\sqrt{3} \;\; \left(\text{equivalent to } y = \pm\sqrt{27}\right) \;\;, \;\; c = 8$$

Answers to Exercise $^\#47$

$$\text{(A)} \; ax < \frac{1}{2} \quad \text{(B)} \; x < \frac{1}{4} \quad \text{(C)} \; x > -\frac{1}{4} \quad \text{(D)} \; x < \frac{1}{2a} \text{ if } a > 0 \;\;, \;\; x > \frac{1}{2a} \text{ if } a < 0$$

$$\text{Note, for example, that } x < \frac{1}{4} \text{ is equivalent to } \frac{1}{4} > x.$$

Answers to Exercise $^\#48$

$$\text{(1)} \; x = -\frac{1}{2} = -0.5 \;\;, \;\; y = -7 \quad \text{(2)} \; x = \frac{9}{2} = 4.5 \;\;, \;\; y = -27$$

Answer to Exercise $^\#49$

$$16x^3 + \frac{9\sqrt{x^7 - 1}}{x^3} = \frac{25(y^3 + 1)^{2/3}}{y^2}$$

Answers to Exercise #50

$$\text{(A) } u^2 + x^2 = 2r^2 + 2n^2 \quad \text{(B) } p = 29 \quad , \quad t = 5 \quad , \quad u = 12\sqrt{5}$$

$$w = 24 \quad , \quad x = 10\sqrt{13} \quad , \quad y = 12 \quad , \quad z = 34$$

Answers to Exercise #51

$$\text{(A) } 2x \quad \text{(B) } 8 \quad \text{(C) } x = -1 \quad , \quad x = 0 \quad , \quad x = 1$$

Answer to Exercise #52

$$w + y = x + z$$

Answers to Exercise #53

$$\text{(1) } x = -\frac{9}{11} \quad \text{(2) } x = 6$$

Answers to Exercise #54

$$\text{(A) } z \geq u \quad \text{(B) } w = \frac{1}{\sqrt{pq}} = \frac{\sqrt{pq}}{pq} \quad \text{(C) } z_{min} = 180 \quad , \quad w = 5$$

$$\text{(D) (1) } w = \frac{5}{2} = 2.5 \quad \text{(2) } w = 10$$

Answer to Exercise #55

$$r = \pm \frac{ap}{y^2} \sqrt{1 + \left(\frac{qx}{py}\right)^2}$$

Answers to Exercise #56

$$\text{(1) } x = 1 \quad , \quad y = -\frac{1}{2} = -0.5 \quad , \quad z = -2$$

$$\text{(2) } x = -4 \quad , \quad y = 2 \quad , \quad z = -2$$

$$\text{(3) } x = \frac{-3 - \sqrt{10}}{2} \quad , \quad y = -3 - \sqrt{10} \quad , \quad z = \frac{1}{2} = 0.5$$

$$\text{(4) } x = \frac{-3 + \sqrt{10}}{2} \quad , \quad y = -3 + \sqrt{10} \quad , \quad z = \frac{1}{2} = 0.5$$

Answers to Exercise #57

$$k = \frac{3}{2} = 1.5 \quad , \quad P = 3x^5 + 16x^3 \quad , \quad Q = x^2 + 4$$

Answers to Exercise #58

$$x = \frac{1}{2} \quad , \quad y = \frac{2}{3} \quad , \quad z = 3$$

Answer to Exercise #59

$$z = c\sqrt{\frac{x^3}{ka}} = \frac{c\sqrt{kax^3}}{ka}$$

Answers to Exercise #60

$$x = 14 \quad , \quad w = -4 \quad , \quad y = 11$$
$$(x = 14 \quad , \quad w = -2 \quad , \quad y = 44 \quad \text{would also work})$$

Answer to Exercise #61

$$x = \frac{25}{64} = 0.390625$$

Answers to Exercise #62

$$(1)\; w = \frac{5}{2} = 2.5 \quad , \quad x = 39 \quad , \quad y = 6 \quad , \quad z = 26$$
$$(2)\; w = -\frac{5}{2} = -2.5 \quad , \quad x = -39 \quad , \quad y = -26 \quad , \quad z = -6$$

Answers to Exercise #63

$$(A)\; y = ax^2 \quad (B)\; y = 27 \quad , \quad p = 3$$

Answers to Exercise #64

$$(A)\; a = -3 \quad , \quad b = 1 \quad , \quad c = 6 \quad , \quad d = 8 \quad , \quad e = 1 \quad , \quad f = -4$$
$$(B)\; x = -6 \quad , \quad x = 4$$

Answers to Exercise #65

$$(1)\ x = -\frac{104}{9} \quad , \quad y = \frac{175}{9}$$
$$(2)\ x = 16 \quad , \quad y = -15$$

Answer to Exercise #66

$$\frac{y}{z} = \frac{x}{w}$$

Answers to Exercise #67

$$0 < w < 72 \quad , \quad x < -4 \text{ or } x > 0 \quad , \quad y > 2 \text{ or } y < 0 \quad , \quad -8 < z < 0$$

Note, for example, that $-8 < z < 0$ is equivalent to $0 > z > -8$.

Answer to Exercise #68

$$u^2 = t^2 + x^2 + 2xy > t^2 + x^2$$

Answers to Exercise #69

(A) 0   (B) 0   (C) yes

Answers to Exercise #70

$$(1)\ x = 15 \quad (2)\ x = \frac{35}{4} = 8.75$$

Answers to Exercise #71

$\left(\frac{3}{2}, -\frac{1}{2}\right)$ and $\left(\frac{13}{2}, \frac{39}{2}\right)$, which are equivalent to $(1.5, -0.5)$ and $(6.5, 19.5)$

Answers to Exercise #72

$$\text{(A) } y^2 k^2 c^4 = w^2 c^2 + k^2 c^4 \quad \text{(B) } \frac{y^2}{y^2 - 1} = \frac{c^2}{u^2}$$

(C) $0 < u < c$ (equivalent to $c > u > 0$)   , $y > 1$ (equivalent to $1 < y$)

Answers to Exercise #73

$$(1)\ x = \frac{1}{2} = 0.5 \quad (2)\ x = 11$$

Answers to Exercise #74

$$x = \frac{7}{4} \quad , \quad y = \frac{4}{3}$$

Answers to Exercise #75

$(A)\ p = k\sqrt{c^2 + \frac{u^4}{w^2}}$ $\left(\text{equivalent to } p = \sqrt{c^2k^2 + \frac{k^2u^4}{w^2}}\right)$ $(B)\ w = \frac{|x|u^2}{c\sqrt{1-x^2}}$

Answers to Exercise #76

(1) $\left(\frac{\sqrt{6}}{2}, \frac{\sqrt{30}}{3}\right)$ (2) $\left(\frac{\sqrt{30}}{2}, \frac{\sqrt{6}}{3}\right)$ (3) $\left(-\frac{\sqrt{6}}{2}, -\frac{\sqrt{30}}{3}\right)$ (4) $\left(-\frac{\sqrt{30}}{2}, -\frac{\sqrt{6}}{3}\right)$

equivalent to (1) $\left(\sqrt{\frac{3}{2}}, 2\sqrt{\frac{5}{6}}\right)$ (2) $\left(\sqrt{\frac{15}{2}}, \frac{2}{\sqrt{6}}\right)$ (3) $\left(-\sqrt{\frac{3}{2}}, -2\sqrt{\frac{5}{6}}\right)$ (4) $\left(-\sqrt{\frac{15}{2}}, -\frac{2}{\sqrt{6}}\right)$

Answers to Exercise #77

$$(1)\ x = 0 \quad (2)\ x = 64$$

Answers to Exercise #78

$(B)\ y = \frac{13}{10} = 1.3$ $(C)\ x = \frac{4 + 15z}{10}$ (equivalent to $10x = 4 + 15z$)

Answers to Exercise #79

$(A)\ a = 3\sqrt{5}\ \left(\text{equivalent to } \frac{15}{\sqrt{5}}\right)$, $b = 3\sqrt{5}\ \left(\text{equivalent to } \frac{15}{\sqrt{5}}\right)$, $c = \sqrt{5}$

$(B)\ y_{max} = 45$ $(C)\ x = 3$

Answers to Exercise #80

$(1)\ x = \frac{10}{27}$, $y = \frac{13}{324}$, $z = \frac{128}{243}$ $(2)\ x = 1$, $y = \frac{1}{4}$, $z = \frac{2}{3}$

Answers to Exercise #81

$$(A)\ P = 10x^3 + 54x^2 + 41x - 16 \quad , \quad Q = 12x^3 + 10x^2 - 8x$$

$$(B)\ \frac{181}{60} \quad (C)\ x = -\frac{4}{3} \quad , \quad x = 0 \quad , \quad x = \frac{1}{2}$$

Answers to Exercise #82

$$(1)\ x = 0 \quad , \quad y = 4 \quad (2)\ x = 6 \quad , \quad y = 12$$

Answers to Exercise #83

$$(A)\ x = \frac{bz}{a+b} \quad , \quad y = -\frac{az}{a+b} \quad (B)\ w = cz^2$$

Answers to Exercise #84

$$x = \frac{49}{4} \quad , \quad y = \frac{64}{9}$$

Answers to Exercise #85

$$(A)\ (x + y + z)\left(\frac{1}{x} + \frac{1}{y} + \frac{1}{z}\right) = \frac{a + b + 3c}{xyz}$$

$$(B)\ a = 150 \quad , \quad b = 138 \quad , \quad c = 36 \quad (C)\ 11$$

Answers to Exercise #86

$$(1)\ \frac{x}{y} = \frac{-3 - \sqrt{29}}{4} \quad (2)\ \frac{x}{y} = \frac{-3 + \sqrt{29}}{4}$$

Answers to Exercise #87

$$(A)\ R = L\left(\sqrt[3]{\frac{3}{4\pi}}\right)\left[\text{equivalent to } L\left(\frac{3}{4\pi}\right)^{\frac{1}{3}}\right] \quad (B)\ \sqrt[3]{\frac{\pi}{6}}\left[\text{equivalent to } \left(\frac{\pi}{6}\right)^{\frac{1}{3}}\right]$$

$$(C)\ S_{\text{sphere}} < S_{\text{cube}}$$

Answers to Exercise #88

$$(A)\ x^2 + y^2 = z^2 \quad (B)\ x = \pm 15 \quad , \quad y = \pm 20$$

Answers to Exercise #89

$$(1)\ x = \frac{3}{2} = 1.5 \quad (2)\ x = 12$$

Answers to Exercise #90

$$(A)\ \frac{u}{t} = \sqrt{\frac{x}{y}} \quad (B)\ t = \frac{\sqrt{axy} + y\sqrt{b}}{\sqrt{a} + \sqrt{b}} \quad , \quad u = \frac{x\sqrt{a} + \sqrt{bxy}}{\sqrt{a} + \sqrt{b}}$$

Answers to Exercise #91

$$\frac{t}{b(y - t)} + \frac{u}{a(u - x)} = \frac{\sqrt{xy}(t - u)}{\sqrt{ab}(u - x)(y - t)} \quad , \quad u - x = \frac{\sqrt{x}(\sqrt{ax} + \sqrt{by}) - x(\sqrt{a} + \sqrt{b})}{\sqrt{a} + \sqrt{b}}$$

$$y - t = \frac{y(\sqrt{a} + \sqrt{b}) - \sqrt{y}(\sqrt{ax} + \sqrt{by})}{\sqrt{a} + \sqrt{b}} \quad , \quad w = ab\left(\frac{\sqrt{y} - \sqrt{x}}{\sqrt{a} + \sqrt{b}}\right)^2$$

Answers to Exercise #92

$$(A)\ x = -\frac{b}{2a} \quad (B)\ y = \frac{b^2 - 4ac}{4a^2} \quad (C)\ x \pm \sqrt{y} = \frac{-b \pm \sqrt{b^2 - 4ac}}{2a}$$

Answers to Exercise #93

$$0 < w < 2 \quad \text{or} \quad w > 3 \quad , \quad -8 < x < -3 \quad \text{or} \quad x > 0$$
$$6 < y < 24 \quad \text{or} \quad y < 0 \quad , \quad -4 < z < 0 \quad \text{or} \quad z < -18$$
These are equivalent to: $2 > w > 0 \quad \text{or} \quad 3 < w \quad , \quad -3 > x > -8 \quad \text{or} \quad 0 < x$
$$24 > y > 6 \quad \text{or} \quad 0 > y \quad , \quad 0 > z > -4 \quad \text{or} \quad -18 > z$$

Answers to Exercise #94

$$(A)\ y = -\frac{3}{4}x + \frac{25}{4} \quad (C)\ m_\perp = \frac{4}{3}$$

Answers to Exercise #95

$$(1)\ x = 1 \quad , \quad y = \frac{9}{10} = 0.9 \quad (2)\ x = \frac{7}{8} = 0.875 \quad , \quad y = \frac{77}{80} = 0.9625$$

Answers to Exercise #96

$$(A)\ p = c - \frac{b^2}{3a} = -\frac{b^2}{3a} + c \quad , \quad q = \frac{2b^3}{27a^2} - \frac{bc}{3a} + d \quad (B)\ z = u^3 + pu$$

Answers to Exercise #97

$$(A)\ x = \frac{3}{8} = 0.375 \quad , \quad y = \frac{1}{8} = 0.125 \quad (B)\ z = \frac{\sqrt{6} - \sqrt{2}}{4}$$

Answer to Exercise #98

$$\frac{1}{y} - \frac{1}{x} = \frac{1 - q}{kc}$$

Answers to Exercise #99

$$(1)\ w = \frac{116}{7} \quad , \quad x = -\frac{30}{11} \quad , \quad y = \frac{10}{7} \quad , \quad z = \frac{58}{11}$$

$$(2)\ w = 6 \quad , \quad x = 4 \quad , \quad y = 12 \quad , \quad z = 12$$

Answers to Exercise #100

$$(A)\ z = \sqrt{\frac{4u^2w^2 - (u^2 + w^2 - x^2)^2}{16}}$$

$$(B)\ z = \sqrt{\frac{(2uw + u^2 + w^2 - x^2)(2uw - u^2 - w^2 + x^2)}{16}}$$

$$(C)\ z = \sqrt{\frac{[(u + w)^2 - x^2][x^2 - (u - w)^2]}{16}}$$

$$(D)\ z = \sqrt{\frac{(u + w + x)(u + w - x)(u + x - w)(w + x - u)}{16}}$$

$$(E)\ z = \sqrt{s(s - x)(s - w)(s - u)}$$

Answers to Exercise #101

$$p = x + y \quad , \quad q = y - x$$

Suggestion

Place a folded sheet of paper here to cover up the hints to problems that you have not yet attempted. This way, you will be less likely to accidentally see hints that you are not yet ready to see.

Hint for Exercise #1: Square both sides of the equation.

Hints for Exercise #2: Make a common denominator of $12x$. After substituting, either make a common denominator of $x - 12$ or multiply both sides by $x - 12$. Put the quadratic equation in standard form.

Hints for Exercise #3: (A) $(a + b)(c + d) = ac + ad + bc + bd$ and $x^m x^n = x^{m+n}$. (B) $x^{\frac{m}{n}} = \left(\sqrt[n]{x}\right)^m$.

Hints for Exercise #4: $m = \frac{y_2 - y_1}{x_2 - x_1}$. Plug x and y for one point into the equation of a line to solve for the y-intercept. Sketch a graph to help visualize the problem.

Hints for Exercise #5: Isolate $x^{3/2}$, then raise both sides of the equation to the power of $\frac{2}{3}$. Use $a^{\frac{m}{n}} = \left(\sqrt[n]{a}\right)^m$ to perform the arithmetic. That is, a fractional exponent is equivalent to first finding a root and then raising that value to a power.

Hints for Exercise #6: Replace z with $x - y$ in the other two equations. Distribute the minus sign: $-a(b - c) = -a(b) - a(-c) = -ab + ac$.

Hint for Exercise #7: $(2x - 3)^2 = 4x^2 - 12x + 9$.

Hints for Exercise #8: Write the given ratio as $\frac{y}{x} = \frac{\sqrt{3}}{2}$. Rationalize the denominator of $\frac{1}{2+\sqrt{3}}$ by multiplying both the numerator and denominator by the conjugate $\left(2 - \sqrt{3}\right)$.

Hints for Exercise #9: $(a - b)^2 = a^2 - 2ab + b^2$ and $(cd)^2 = c^2 d^2$.

Hint for Exercise #10: Multiply by 2 on both sides of the top equation.

Hints for Exercise #11: Make a common denominator to add the fractions before taking the reciprocal. Recall that $y^{-1} = \frac{1}{y}$.

Hint for Exercise #12: Take the reciprocal of each side of each given equation.

Hints for Exercise #13: The subscripts simply help to keep track of the initial and final values. As an example of working with subscripts, consider the formula $c = \frac{a+b}{2}$. For the initial values, $c_1 = \frac{a_1+b_1}{2}$, and for the final values, $c_2 = \frac{a_2+b_2}{2}$. If a quantity is constant, its initial and final values are equal. For example, if a is constant, $a_1 = a_2$.

Hints for Exercise #14: First square each given expression. Then find the third power.

Hints for Exercise #15: (A)/(B) First multiply both sides of $z = \frac{x}{y}$ by y.

(C) Solve for x^2 in terms of w and z. When taking the square root of both sides, there are both positive and negative roots. For example, $(-2)^2 = 4$ and $2^2 = 4$. Be careful with the direction of the inequality. Similarly, solve for y^2 in terms of w and z.

Hint for Exercise #16: First find z, then y, then x, and then w.

Hints for Exercise #17: (A) Multiply the first term by $\frac{\sqrt{x}-\sqrt{3}}{\sqrt{x}-\sqrt{3}}$ and the second term by $\frac{\sqrt{x}+\sqrt{3}}{\sqrt{x}+\sqrt{3}}$. (B) Cross multiply: $\frac{a}{b} = \frac{c}{d}$ becomes $ad = bc$.

Hints for Exercise #18: Either square both sides of the equation or let $y = \sqrt{x}$. In the latter case, use $y^2 = x$ to find x after solving for y. Put the quadratic equation in standard form.

Hints for Exercise #19: (C) First solve for y in terms of u, w, and z. The formulas for x and y should suggest what to do next.

Hint for Exercise #20: $(a + b)(c - d) = ac - ad + bc - bd$.

Hints for Exercise #21: Make a common denominator of $(x - 3)x$. Distribute the minus sign. For example, $-a(b - c) = -a(b) - a(-c) = -ab + ac$.

Hint for Exercise #22: Cross multiply: $\frac{a}{b} = \frac{c}{d}$ becomes $ad = bc$.

Hints for Exercise #23: (A) Since $w > 0$, it follows that $2w > 0$ and that $2w + 10 > 10$, such that $u > 10$. (B) First write $D = 2x$. Isolate x in this simple equation and then substitute. Surprisingly, perhaps, many students make a mistake with this simple substitution. Students who take the time and care to do this one clear step at a time are more likely to get it right. (C) Isolate $\frac{5x}{2}$ in the top given equation and consider for a moment about why this should be helpful.

Hints for Exercise #24: Recall the rules for working with exponents, such as $(a^b)^c = a^{bc}$, $(pq)^r = p^r q^r$, $\sqrt{uw} = \sqrt{u}\sqrt{w}$, $\sqrt{t^n} = t^{n/2}$, $u^m u^n = u^{m+n}$, and $\left(\frac{p}{q}\right)^{-k} = \left(\frac{q}{p}\right)^k = \frac{q^k}{p^k}$.

Hints for Exercise #25: Distribute the minus sign. For example, $-a(b - c) = -a(b) - a(-c) = -ab + ac$. One variable should be relatively easy to determine.

Hints for Exercise #26: Cross multiply: $\frac{a}{b} = \frac{c}{d}$ becomes $ad = bc$. Put the quadratic equation in standard form.

Hints for Exercise #27: Set one pair of equations equal to solve for x, and then plug x into one of the equations to find y. Do this for all three possible pairs. Sketch a graph to help visualize the problem and to check that each answer makes sense.

Hints for Exercise #28: Multiply the first term by $\frac{\sqrt{x^2+1}}{\sqrt{x^2+1}}$. To rationalize the denominator, multiply by the conjugate. The conjugate of an expression of the form $t\sqrt{u} + w$ is $-t\sqrt{u} + w$. Recall that $\sqrt{y}\sqrt{y} = y$, $\frac{\sqrt{y}}{y} = \frac{1}{\sqrt{y}}$, and $\sqrt{y} = y^{1/2}$.

Hints for Exercise #29: (B) Use $2x = wz$ and an answer to Part A. (C) The problem states that w and z are each positive. Use the answer to Part B.

Hints for Exercise #30: (A) Expand $(ax + b)^2$. Recall that $(tu)^n = t^n u^n$. Reason out what a needs to be in order to match the coefficient of x^2. Knowing a, now solve for the value of b needed to match the coefficient of the linear term. Finally, determine the value of c needed to make the two equations identical. (B) Follow the same strategy as Part A using symbols instead of numbers.

Hint for Exercise #31: Isolate p and q in $u = \frac{x}{p}$ and $w = \frac{x}{q}$.

Hints for Exercise #32: Take the reciprocal of both sides, isolate the next set of parentheses, and repeat.

Hints for Exercise #33: Multiply out the expression on the left. Factor out coefficients of like terms (such as the x^2 terms). The coefficients must match on the two sides.

Hints for Exercise #34: $\sqrt{x}\sqrt{x} = x$, $\frac{\sqrt{x}}{\sqrt{x}} = 1$, $\frac{x}{\sqrt{x}} = \sqrt{x}$, $\frac{\sqrt{x}}{x} = \frac{1}{\sqrt{x}}$, and $\frac{1}{x\sqrt{x}} = \frac{\sqrt{x}}{x^2}$. At the end, first factor out $\sqrt{x}$ and then factor out $\frac{1}{x^2}$.

Hints for Exercise #35: Multiply both sides of each equation by the denominator. Divide the top two equations. Multiply this new equation by the third equation. When you square root both sides, consider both positive and negative roots.

Hints for Exercise #36: Multiply the numerator and denominator each by $p^{\frac{1-k}{k}}$. Distribute in the numerator, but not the denominator. Multiply the numerator and denominator each by $q^{\frac{1-k}{k}}$. Distribute in the denominator, but not the numerator. Look for significant cancellation. Apply the rule $\left(\frac{u}{w}\right)^m = \frac{u^m}{w^m}$.

Hints for Exercise #37: (A) Multiply both sides of each equation by the denominator. (B) Set $z = 0$ in the answer to Part A.

Hints for Exercise #38: $m = \frac{y_2 - y_1}{x_2 - x_1}$. Plug x and y for one point into the equation of a line to solve for the y-intercept. $m_\perp = -\frac{1}{m}$. The coordinates of the midpoint can be found by averaging the x-coordinates and the y-coordinates of the given points.

Hints for Exercise #39: (A) Put the quadratic equation in standard form. (C) $\frac{y}{x} = 15$.

Hints for Exercise #40: Substitute q into the top equation. Square the bottom equation and plug it into the top equation. Combine like terms. Look for cancellation.

Hints for Exercise #41: Cross multiply: $\frac{a}{b} = \frac{c}{d}$ becomes $ad = bc$. It is possible to avoid the quadratic formula by square rooting both sides at the appropriate moment, but remember to allow for both positive and negative roots.

Hints for Exercise #42: Factor the numerator and denominator, and then factor the quadratic expression.

Hints for Exercise #43: $(a - b)^2 = a^2 - 2ab + b^2$. Put the quadratic equation in standard form.

Hints for Exercise #44: Recall the rules for working with exponents, such as $(t^a)^b = t^{ab}$, $t^m t^n = t^{m+n}$, $\frac{t^m}{t^n} = t^{m-n}$, and $(x^4)^{1/4} = x^1 = x$. Consider both positive and negative roots. Think through the possible combinations of signs.

Hints for Exercise #45: (A)-(D) First multiply the left expression by 1, and then multiply it by $-x$. (E) Use the pattern to generalize the formula to an expression ending with x^n. Divide both sides by $(1 - x)$.

Hints for Exercise #46: $(a + b)^2 = a^2 + 2ab + b^2$ and $(x + c)(x - c) = x^2 - c^2$. Recall that $\sqrt{u}\sqrt{u} = u$ and $\sqrt{p}\sqrt{q} = \sqrt{pq}$. At the end, divide both sides by $a^2 b^2$.

Hint for Exercise #47: If you multiply or divide both sides of an inequality by a negative number, this reverses the direction of the inequality.

Hints for Exercise #48: Isolate y in the second equation. Put the quadratic equation in standard form.

Hints for Exercise #49: Factor the square root and the right side of the equation. Then figure out what to divide by on both sides of the equation.

Hints for Exercise #50: (A) Add the top two equations. Express w, y, and z in terms of n and r. (B) First find y and t.

Hints for Exercise #51: Make common denominators of $(x + 1)x$ and $(x - 1)x$. Simplify as much as possible before taking the reciprocals.

Hints for Exercise #52: Isolate $\frac{a}{b}$ in each equation. Recall that $(p + q)(p - q) = p^2 - q^2$. Compare the equations closely. What conclusion can you draw?

Hints for Exercise #53: Cross multiply: $\frac{a}{b} = \frac{c}{d}$ becomes $ad = bc$. Put the quadratic equation in standard form.

Hints for Exercise #54: (A)-(C) Any real number squared is nonnegative. (D) Square both sides of the equation for z. Put the quadratic equation in standard form.

Hints for Exercise #55: Plug in the expression for z. Apply the rule $(b^c)^d = b^{cd}$. Distribute and factor. Square root both sides of the equation.

Hints for Exercise #56: Plug the expression for x into the equation for y. Put the quadratic equation in standard form.

Hints for Exercise #57: $u^{1/2} = \sqrt{u}, u^{-1/2} = \frac{1}{\sqrt{u}}$, and $\frac{u^{-1/2}}{u} = \frac{1}{u^{3/2}}$. Factor out $(x^2 + 4)^{-1/2}$.

Hints for Exercise #58: Isolate x in the top equation. Take the reciprocal of both sides. Plug this into the bottom equation. Isolate z. Make a common denominator. Plug this into the middle equation. Put the quadratic equation in standard form.

Hints for Exercise #59: Combine the top two equations first. After using the bottom equation, cross multiply: $\frac{a}{b} = \frac{c}{d}$ becomes $ad = bc$.

Hint for Exercise #60: Square both sides of the equation: $(p - q)^2 = p^2 - 2pq + q^2$.

Hints for Exercise #61: Multiply both sides by $\sqrt{x}$. Recall that $\sqrt{x}\sqrt{x} = x$ and $\frac{\sqrt{x}}{\sqrt{x}} = 1$.

Hints for Exercise #62: Multiply the top two equations. After using all of the equations, divide both sides by 78. When taking the square root of both sides, allow for both positive and negative roots. For example, $(-2)^2 = 4$ and $2^2 = 4$.

Hints for Exercise #63: Square both sides of the equation. After squaring, the absolute values may be removed.

Hint for Exercise #64: Factor $x^2 + 2x - 24$.

Hints for Exercise #65: Isolate one variable in the top equation and substitute. Put the quadratic equation in standard form.

Hints for Exercise #66: Divide the two equations that have exponents. Isolate $\frac{w}{x}$ and $\frac{r}{s}$ in the top equations and use these ratios.

Hints for Exercise #67: What happens to an inequality when you cross multiply? If you don't know, try simple examples like $\frac{1}{f} > \frac{1}{2}$ and $\frac{1}{g} > -\frac{1}{3}$. Test it with numbers.

Hints for Exercise #68: Derive an equation of the form $u^2 = t^2 + x^2 + \cdots$, where the three dots indicate that there is at least one more term. Show that the expression indicated by the three dots is positive.

Hint for Exercise #69: Distribute and look for like terms to combine.

Hints for Exercise #70: Expand both sides. The cubic terms should cancel. Put the quadratic equation in standard form.

Hints for Exercise #71: Set the equations equal. Put the quadratic equation in standard form.

Hints for Exercise #72: Square both sides of the second equation. Make a common denominator.

Hints for Exercise #73: Square both sides of the equation, isolate the next square root, and repeat. Put the quadratic equation in standard form.

Hint for Exercise #74: Multiply both sides of the first equation by 12 and the second equation by 4.

Hints for Exercise #75: (A) Square both sides of two of the equations and add them together. (B) Divide two of the equations.

Hints for Exercise #76: Square both sides of the equation for the hyperbola. Let $u = x^2$. Put the quadratic equation in standard form. Use $\pm\sqrt{u} = x$ to find x.

Hints for Exercise #77: $(a - b)^2 = a^2 - 2ab + b^2$. Recall that $\sqrt{x}\sqrt{x} = x$. Factor out $\sqrt{x}$.

Hints for Exercise #78: It is possible to isolate $3z - 2x$ (or any multiple of this) in each equation. Let $w = 3z - 2x$ to get two equations with w and y.

Hints for Exercise #79: (A) Recall that $(t - u)^2 = t^2 - 2tu + u^2$ and $(rs)^2 = r^2s^2$. Distribute the minus sign: $-(-p) = p$. The coefficients of the quadratic term, linear term, and constant term must be equal. Use this to solve for the constants. (B) Any real number squared is nonnegative.

Hints for Exercise #80: Isolate z in the top equation and y in the middle equation. Substitute these into the bottom equation.

Hints for Exercise #81: Multiply the first term by $\frac{6x+8}{6x+8}$ and the second term by $\frac{2x-1}{2x-1}$. After simplifying, make a common denominator with the last term.

Hints for Exercise #82: Square both sides of the top equation. Make a substitution. Put the quadratic equation in standard form.

Hint for Exercise #83: The top two equations are not needed until Part B.

Hints for Exercise #84: This time, it is not helpful to square an equation. To avoid the square roots, let $t = \sqrt{x}$ and $u = \sqrt{y}$. After finding t and u, use $t^2 = x$ and $u^2 = y$.

Hints for Exercise #85: Multiply $\frac{1}{x} + \frac{1}{y} + \frac{1}{z}$ by xyz and simplify. Now divide both sides by xyz and multiply both sides by $(x + y + z)$. Compare with the given equations.

Hints for Exercise #86: Add an expression to both sides that completes the square (like the concept featured in Exercise 30). Square root both sides.

Hints for Exercise #87: (A) Set the expressions for the volumes equal, isolate R^3, and take the cube root of both sides. (B) Divide the expressions for the surface areas and plug in the answer to Part A. The final answer will be fairly simple if you are fluent in your rules for working with exponents.

Hints for Exercise #88: A proportion of the form $p{:}q = r{:}s$ can be expressed as $\frac{p}{q} = \frac{r}{s}$. Cross multiply: $\frac{a}{b} = \frac{c}{d}$ becomes $ad = bc$.

Hints for Exercise #89: Raise both sides to the power of 3, isolate the next set of parentheses, raise both sides to the power of $\frac{1}{4}$, and so on. Only the positive roots lead to real answers in this problem.

Hints for Exercise #90: (A) In the top equation, make the middle term and the last term have a common denominator. In the bottom equation, make the first term and the last term have a common denominator. Combine these results and cross multiply. (B) Inspect the two given equations closely. It should be obvious that two expressions must be equal. Set these equal. Simplify and cross multiply. Substitute using the answer to Part A.

Hints for Exercise #91: Use the first hint to Part B of Exercise 90 for the top equation. Use the answers to Part B of Exercise 90 for the middle equations. Use the top three equations and the given equation to derive the last equation.

Hints for Exercise #92: (A) Factor out $\sqrt{y}$. (B) Use the answer to Part A.

Hints for Exercise #93: Work out all possible cases for the signs of the denominators. Zero, one, or both denominators may be negative for different intervals.

Hints for Exercise #94: First find the equation of the radius that connects the center of the circle to the point of intersection. Then use $m = -\dfrac{1}{m_\perp}$.

Hint for Exercise #95: Remove the absolute values and write $\pm$.

Hints for Exercise #96: Multiply the expressions out. Combine like terms. Compare equations to identify the constants.

Hints for Exercise #97: A straightforward method is to square both sides. Compare rational and irrational parts to make two equations. (There is also a clever method.)

Hints for Exercise #98: Isolate the square root, and then square both sides. Isolate zs in the top equation, square both sides of the top two equations, and add these squared equations together.

Hints for Exercise #99: Isolate one reciprocal of a variable in each of the top two equations, make a common denominator, and take the reciprocal of both sides.

Hints for Exercise #100: (A) Write $z = \sqrt{\dfrac{w^2 t^2}{4}} = \sqrt{\dfrac{4w^2 t^2}{16}}$.
(B)-(D) $f^2 - g^2 = (f + g)(f - g)$.

Hints for Exercise #101: Multiply the expressions out in both equations and compare. Set the real parts and the imaginary parts equal to get two equations to solve.

Solution to Exercise #1	
(A) $1 + \sqrt{x} = \sqrt{4 + x}$ $\left(1 + \sqrt{x}\right)^2 = \left(\sqrt{4 + x}\right)^2$ $1 + 2\sqrt{x} + x = 4 + x$ $2\sqrt{x} = 3$ $\sqrt{x} = \dfrac{3}{2}$ $x = \left(\dfrac{3}{2}\right)^2 = \boxed{\dfrac{9}{4}} = \boxed{2.25}$	(B) $1 + \sqrt{\dfrac{9}{4}} = \sqrt{4 + \dfrac{9}{4}}$ $1 + \dfrac{3}{2} = \sqrt{\dfrac{16}{4} + \dfrac{9}{4}}$ $\dfrac{2}{2} + \dfrac{3}{2} = \sqrt{\dfrac{25}{4}}$ $\dfrac{5}{2} = \dfrac{5}{2}$ ✔

Solution to Exercise #2	
(A) $\dfrac{1}{y} = \dfrac{1}{12} - \dfrac{1}{x}$	Make a common denominator of $12x$.
$\dfrac{1}{y} = \dfrac{x}{12x} - \dfrac{12}{12x} = \dfrac{x - 12}{12x}$	Take the reciprocal of both sides.
$y = \dfrac{12x}{x - 12}$	Replace y with $\dfrac{12x}{x-12}$ in $x + y = 49$.
$x + \dfrac{12x}{x - 12} = 49$	Make a common denominator of $x - 12$.
$\dfrac{x(x - 12)}{x - 12} + \dfrac{12x}{x - 12} = 49$	Distribute: $a(b - c) = ab - ac$.
$\dfrac{x^2 - 12x}{x - 12} + \dfrac{12x}{x - 12} = 49$	
$\dfrac{x^2 - 12x + 12x}{x - 12} = 49$	$12x$ cancels out.
$\dfrac{x^2}{x - 12} = 49$	Multiply both sides by $x - 12$.
$x^2 = 49(x - 12) = 49x - 588$	Put the quadratic equation in standard
$x^2 - 49x + 588 = 0$	form. Bring $49x$ and 588 to the left.
$a = 1, b = -49, c = 588$	Use the quadratic formula (or factor).

$$x = \frac{-b \pm \sqrt{b^2 - 4ac}}{2a}$$

$$x = \frac{-(-49) \pm \sqrt{(-49)^2 - 4(1)(588)}}{2(1)}$$

$$x = \frac{49 \pm \sqrt{2401 - 2352}}{2}$$

$$x = \frac{49 \pm \sqrt{49}}{2} = \frac{49 \pm 7}{2}$$

$$x = \frac{49 - 7}{2} = \frac{42}{2} = \boxed{21}$$

$$\text{or } x = \frac{49 + 7}{2} = \frac{56}{2} = \boxed{28}$$

$$y = 49 - x = 49 - 21 = \boxed{28}$$

$$\text{or } y = 49 - 28 = \boxed{21}$$

(B) $\dfrac{1}{21} + \dfrac{1}{28} = \dfrac{4}{84} + \dfrac{3}{84} = \dfrac{7}{84} = \dfrac{1}{12}$ ✓

$$21 + 28 = 49 \quad ✓$$

An alternative to using the quadratic formula is to factor $x^2 - 49x + 588$ as $(x - 21)(x - 28)$.

Since $(49)(12) = 588$, the arithmetic could actually be simplified somewhat by factoring 49 inside of the square root.

There are two possible solutions for x: one for each possible sign of the $\pm$. Work out the answer for each case.

One solution is $x = 28$ and $y = 21$.

Another solution is $x = 21$ and $y = 28$.

Solution to Exercise #3

(A) $\dfrac{\left(x^{\frac{3}{4}} + x^{\frac{1}{4}}\right)\left(x^{\frac{2}{3}} - x^{\frac{1}{6}}\right)}{x^{\frac{5}{12}}}$

$$= \frac{x^{\frac{3}{4}}x^{\frac{2}{3}} - x^{\frac{3}{4}}x^{\frac{1}{6}} + x^{\frac{1}{4}}x^{\frac{2}{3}} - x^{\frac{1}{4}}x^{\frac{1}{6}}}{x^{\frac{5}{12}}}$$

$$= \frac{x^{\frac{3}{4}+\frac{2}{3}} - x^{\frac{3}{4}+\frac{1}{6}} + x^{\frac{1}{4}+\frac{2}{3}} - x^{\frac{1}{4}+\frac{1}{6}}}{x^{\frac{5}{12}}}$$

$$= \frac{x^{\frac{9}{12}+\frac{8}{12}} - x^{\frac{9}{12}+\frac{2}{12}} + x^{\frac{3}{12}+\frac{8}{12}} - x^{\frac{3}{12}+\frac{2}{12}}}{x^{\frac{5}{12}}}$$

$$= \frac{x^{\frac{17}{12}} - x^{\frac{11}{12}} + x^{\frac{11}{12}} - x^{\frac{5}{12}}}{x^{\frac{5}{12}}}$$

Apply the "foil" method:

$$(a + b)(c + d)$$
$$= ac + ad + bc + bd$$

Recall that $x^m x^n = x^{m+n}$.

Make a common denominator to add the fractions.

$$\frac{3}{4} = \frac{9}{12}$$

$$\frac{2}{3} = \frac{8}{12}$$

$$\frac{1}{4} = \frac{3}{12}$$

$$\frac{1}{6} = \frac{2}{12}$$

$$= \frac{x^{\frac{17}{12}} - x^{\frac{5}{12}}}{x^{\frac{5}{12}}}$$

$$= \frac{x^{\frac{17}{12}}}{x^{\frac{5}{12}}} - \frac{x^{\frac{5}{12}}}{x^{\frac{5}{12}}} = x^{\frac{17}{12} - \frac{5}{12}} - x^{\frac{5}{12} - \frac{5}{12}}$$

$$= x^{\frac{12}{12}} - x^0 = x^1 - x^0 = \boxed{x - 1}$$

(B) $x - 1 = 4096 - 1 = \boxed{4095}$

$$\frac{\left(4096^{\frac{3}{4}} + 4096^{\frac{1}{4}}\right)\left(4096^{\frac{2}{3}} - 4096^{\frac{1}{6}}\right)}{4096^{\frac{5}{12}}}$$

$$= \frac{\left[\left(\sqrt[4]{4096}\right)^3 + \sqrt[4]{4096}\right]\left[\left(\sqrt[3]{4096}\right)^2 - \sqrt[6]{4096}\right]}{\left(\sqrt[12]{4096}\right)^5}$$

$$= \frac{[(8)^3 + 8][(16)^2 - 4]}{(2)^5}$$

$$= \frac{(512 + 8)(256 - 4)}{32}$$

$$= \frac{(520)(252)}{(8)(4)} = \left(\frac{520}{8}\right)\left(\frac{252}{4}\right) = (65)(63)$$

$$= \boxed{4095} \quad \checkmark$$

$x^{\frac{11}{12}}$ cancels out.

Apply the distributive property:

$$\frac{a - b}{c} = \frac{a}{c} - \frac{b}{c}$$

Recall that $\dfrac{x^m}{x^n} = x^{m-n}$.

Recall that $x^1 = x$ and $x^0 = 1$.

$$x^{\frac{m}{n}} = \left(\sqrt[n]{x}\right)^m$$

$$4096^{\frac{3}{4}} = \left(\sqrt[4]{4096}\right)^3$$

$$4096^{\frac{1}{4}} = \sqrt[4]{4096}$$

$$4096^{\frac{2}{3}} = \left(\sqrt[3]{4096}\right)^2$$

$$4096^{\frac{1}{6}} = \sqrt[6]{4096}$$

$$2^{12} = 4096$$

$$4^6 = 4096$$

$$8^4 = 4096$$

$$16^3 = 4096$$

Solution to Exercise #4

(A) $m_1 = \dfrac{8 - (-4)}{-4 - 4} = \dfrac{8 + 4}{-8} = \dfrac{12}{-8} = -\dfrac{3}{2}$

$$y = m_1 x + b_1 = -\frac{3}{2}x + b_1$$

$$8 = -\frac{3}{2}(-4) + b_1 = 3(2) + b_1 = 6 + b_1$$

$$8 - 6 = 2 = b_1$$

$$\boxed{y = -\frac{3}{2}x + 2}$$

$$m_2 = \frac{-10 - (-4)}{-8 - 4} = \frac{-10 + 4}{-12} = \frac{-6}{-12} = \frac{1}{2}$$

$$m = \frac{y_2 - y_1}{x_2 - x_1}$$

For $(-4, 8)$, $x_2 = -4$ and $y_2 = 8$.

For $(4, -4)$, $x_1 = 4$ and $y_1 = -4$.

Plug $m_1 = -\frac{3}{2}$ into $y = m_1 x + b_1$.

To solve for b_1, plug in $x = -4$ and $y = 8$ for the point $(-4, 8)$.

Alternatively, plug in $x = 4$ and $y = -4$ for the point $(4, -4)$.

For $(-8, -10)$, $x_2 = -10$ and $y_2 = -8$.

$$y = m_2 x + b_2 = \frac{x}{2} + b_2$$

$$-10 = \frac{1}{2}(-8) + b_2 = -4 + b_2$$

$$-10 + 4 = -6 = b_2$$

$$\boxed{y = \frac{x}{2} - 6}$$

$$m_3 = \frac{-10 - 8}{-8 - (-4)} = \frac{-18}{-8 + 4} = \frac{-18}{-4} = \frac{9}{2}$$

$$y = m_3 x + b_3 = \frac{9}{2}x + b_3$$

$$-10 = \frac{9}{2}(-8) + b_3 = 9(-4) + b_3$$

$$= -36 + b_3$$

$$-10 + 36 = 26 = b_3$$

$$\boxed{y = \frac{9}{2}x + 26}$$

$$(B) \ -4 = -\frac{3}{2}(4) + 2 = -3(2) + 2$$

$$= -6 + 2 = -4 \quad \checkmark$$

$$8 = -\frac{3}{2}(-4) + 2 = -3(-2) + 2$$

$$= 6 + 2 = 8 \quad \checkmark$$

$$-4 = \frac{4}{2} - 6 = 2 - 6 = -4 \quad \checkmark$$

$$-10 = \frac{-8}{2} - 6 = -4 - 6 = -10 \quad \checkmark$$

$$8 = \frac{9}{2}(-4) + 26 = 9(-2) + 26$$

$$= -18 + 26 = 8 \quad \checkmark$$

$$-10 = \frac{9}{2}(-8) + 26 = 9(-4) + 26$$

$$= -36 + 26 = -10 \quad \checkmark$$

For $(4, -4)$, $x_1 = 4$ and $y_1 = -4$.

Plug $m_2 = \frac{1}{2}$ into $y = m_2 x + b_2$.

To solve for b_2, plug in $x = -8$ and $y = -10$ for the point $(-8, -10)$.

Alternatively, plug in $x = 4$ and $y = -4$ for the point $(4, -4)$.

For $(-8, -10)$, $x_2 = -10$ and $y_2 = -8$.

For $(-4, 8)$, $x_1 = -4$ and $y_1 = 8$.

Plug $m_3 = \frac{9}{2}$ into $y = m_3 x + b_3$.

To solve for b_3, plug in $x = -8$ and $y = -10$ for the point $(-8, -10)$.

Alternatively, plug in $x = -4$ and $y = 8$ for the point $(-4, 8)$.

Put $x = 4$ and $y = -4$ into $y = -\frac{3}{2}x + 2$.

Put $x = -4$ and $y = 8$ into $y = -\frac{3}{2}x + 2$.

Put $x = 4$ and $y = -4$ into $y = \frac{x}{2} - 6$.

Put $x = -8$ and $y = -10$ into $y = \frac{x}{2} - 6$.

Put $x = -4$ and $y = 8$ into $y = \frac{9}{2}x + 26$.

Put $x = -8$ and $y = -10$ into $y = \frac{9}{2}x + 26$.

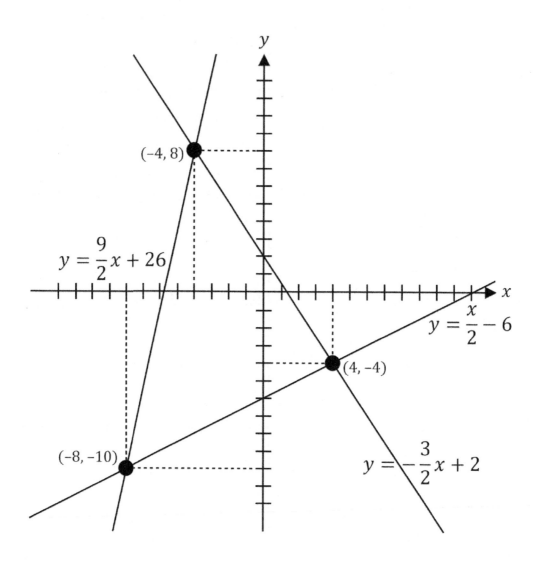

Solution to Exercise #5

(A) $25 = \dfrac{11}{3} + \dfrac{9}{x^{3/2}}$

$25 - \dfrac{11}{3} = \dfrac{9}{x^{3/2}}$

$\dfrac{75}{3} - \dfrac{11}{3} = \dfrac{9}{x^{3/2}}$

$\dfrac{64}{3} = \dfrac{9}{x^{3/2}}$

$64x^{3/2} = 3(9)$

$64x^{3/2} = 27$

$x^{3/2} = \dfrac{27}{64}$

$\left(x^{\frac{3}{2}}\right)^{\frac{2}{3}} = \left(\dfrac{27}{64}\right)^{\frac{2}{3}}$

$x = \left(\dfrac{27}{64}\right)^{\frac{2}{3}} = \dfrac{27^{2/3}}{64^{2/3}} = \dfrac{\left(\sqrt[3]{27}\right)^2}{\left(\sqrt[3]{64}\right)^2} = \dfrac{3^2}{4^2}$

$= \boxed{\dfrac{9}{16}} = \boxed{0.5625}$

(B) $x^{3/2} = a$

$\left(x^{\frac{3}{2}}\right)^{\frac{2}{3}} = a^{\frac{2}{3}}$

$a^{2/3} = \left(\sqrt[3]{a}\right)^2$

(C) $\left(\dfrac{9}{16}\right)^{\frac{3}{2}} = \dfrac{9^{3/2}}{16^{3/2}} = \dfrac{\left(\sqrt{9}\right)^3}{\left(\sqrt{16}\right)^3} = \dfrac{3^3}{4^3} = \dfrac{27}{64}$

$25 - \dfrac{9}{\dfrac{27}{64}} = 25 - 9 \times \dfrac{64}{27}$

$= 25 - \dfrac{64}{3} = \dfrac{75}{3} - \dfrac{64}{3} = \dfrac{11}{3} \quad \checkmark$

Solution to Exercise #6

(A) $(14x + 22x) + (21 - 33) - 36z = 0$

$36x - 12 - 36z = 0$

$36x - 12 - 36(x - y) = 0$

$36x - 12 - 36x + 36y = 0$

$-12 + 36y = 0$

$36y = 12$

$y = \dfrac{12}{36} = \boxed{\dfrac{1}{3}}$

$(54y - 24y) + (5 - 21) + 36z = 0$

$30y - 16 + 36z = 0$

(B) $14\left(\dfrac{1}{2}\right) + 21 - 36\left(\dfrac{1}{6}\right) - 33 + 22\left(\dfrac{1}{2}\right)$

$= 0$

$7 + 21 - 6 - 33 + 11 = 39 - 39 = 0 \quad \checkmark$

$54\left(\dfrac{1}{3}\right) - 21 + 36\left(\dfrac{1}{6}\right) + 5 - 24\left(\dfrac{1}{3}\right) = 0$

$18 - 21 + 6 + 5 - 8 = 29 - 29 = 0 \quad \checkmark$

$\dfrac{1}{2} - \dfrac{1}{3} = \dfrac{1}{6} \quad \checkmark$

$$30\left(\frac{1}{3}\right) - 16 + 36z = 0$$

$$10 - 16 + 36z = 0$$

$$-6 + 36z = 0$$

$$36z = 6$$

$$z = \frac{6}{36} = \boxed{\frac{1}{6}}$$

$$x - y = z$$

$$x = y + z = \frac{1}{3} + \frac{1}{6} = \frac{2}{6} + \frac{1}{6} = \frac{3}{6} = \boxed{\frac{1}{2}}$$

Solution to Exercise #7

$$(A)\ (4x^2 - 12x + 9)\left(8x^2 - 6x + \frac{3x + 4}{2x - 3}\right)$$

$$= (4x^2 - 12x + 9)(8x^2) + (4x^2 - 12x + 9)(-6x) + \frac{(4x^2 - 12x + 9)(3x + 4)}{2x - 3}$$

$$= 32x^4 - 96x^3 + 72x^2 - 24x^3 + 72x^2 - 54x + \frac{(2x - 3)^2(3x + 4)}{2x - 3}$$

$$= 32x^4 - 120x^3 + 144x^2 - 54x + (2x - 3)(3x + 4)$$

$$= 32x^4 - 120x^3 + 144x^2 - 54x + 6x^2 + 8x - 9x - 12$$

$$= \boxed{32x^4 - 120x^3 + 150x^2 - 55x - 12}$$

The "trick" to fully simplifying Part A is to recognize that $(2x - 3)^2 = (4x^2 - 12x + 9)$, such that $\frac{4x^2 - 12x + 9}{2x - 3} = \frac{(2x - 3)^2}{2x - 3} = 2x - 3$.

$$(B)\ 32(2)^4 - 120(2)^3 + 150(2)^2 - 55(2) - 12$$

$$= 32(16) - 120(8) + 150(4) - 110 - 12$$

$$= 512 - 960 + 600 - 122 = \boxed{30}$$

$$[4(2)^2 - 12(2) + 9]\left[8(2)^2 - 6(2) + \frac{3(2) + 4}{2(2) - 3}\right]$$

$$= [4(4) - 24 + 9]\left[8(4) - 12 + \frac{6 + 4}{4 - 3}\right]$$

$$= (16 - 15)\left[32 - 12 + \frac{10}{1}\right]$$

$$= (1)(20 + 10) = (1)(30) = \boxed{30} \quad \checkmark$$

(C) $2x - 3 = 0$ $$2x = 3$$ $$x = \boxed{\frac{3}{2}} = \boxed{1.5}$$	If $2x - 3$ equals zero, the fraction will involve division by zero. In this case, the fraction is zero divided by zero, which is indeterminate.

Solution to Exercise #8

(A) $\dfrac{y}{x} = \dfrac{\sqrt{3}}{2}$	$y{:}x$ equals $\dfrac{y}{x}$. A ratio is a fraction.
$y = \dfrac{x\sqrt{3}}{2}$	Multiply both sides by x.
$x + y = \dfrac{1}{2}$	Replace y with $\dfrac{x\sqrt{3}}{2}$.
$x + \dfrac{x\sqrt{3}}{2} = \dfrac{1}{2}$	Multiply both sides by 2.
$2x + x\sqrt{3} = 1$	Factor out the x.
$x(2 + \sqrt{3}) = 1$	Divide both sides by $2 + \sqrt{3}$.
$x = \dfrac{1}{2 + \sqrt{3}} = \dfrac{1}{2 + \sqrt{3}}\left(\dfrac{2 - \sqrt{3}}{2 - \sqrt{3}}\right)$	Multiply the numerator and denominator each by $2 - \sqrt{3}$.
$x = \dfrac{2 - \sqrt{3}}{4 - 3} = \dfrac{2 - \sqrt{3}}{1} = \boxed{2 - \sqrt{3}}$	$(2 + \sqrt{3})(2 - \sqrt{3})$ $= 2(2) - 2\sqrt{3} + 2\sqrt{3} - \sqrt{3}\sqrt{3}$ $= 4 - 3 = 1$
$y = \dfrac{x\sqrt{3}}{2} = (2 - \sqrt{3})\dfrac{\sqrt{3}}{2} = \boxed{\sqrt{3} - \dfrac{3}{2}}$ $= \boxed{\dfrac{2\sqrt{3} - 3}{2}}$	$(2 - \sqrt{3})\dfrac{\sqrt{3}}{2} = 2\dfrac{\sqrt{3}}{2} - \sqrt{3}\dfrac{\sqrt{3}}{2} = \sqrt{3} - \dfrac{3}{2}$
(B) $\dfrac{y}{x} = \dfrac{2\sqrt{3} - 3}{2(2 - \sqrt{3})} = \dfrac{(2 - \sqrt{3})\sqrt{3}}{2(2 - \sqrt{3})}$ $= \dfrac{\sqrt{3}}{2} \quad \checkmark$	Factor: $2\sqrt{3} - 3 = (2 - \sqrt{3})\sqrt{3}$.
$x + y = 2 - \sqrt{3} + \sqrt{3} - \dfrac{3}{2} = \dfrac{4}{2} - \dfrac{3}{2} = \dfrac{1}{2} \quad \checkmark$	

Solution to Exercise #9

$$(A)\ f = (xq - yp)^2 + (xr - zp)^2 + (yr - zq)^2$$

$$f = \boxed{x^2q^2 - 2xypq + y^2p^2 + x^2r^2 - 2xzpr + z^2p^2 + y^2r^2 - 2yzqr + z^2q^2}$$

$$g = (x^2 + y^2 + z^2)(p^2 + q^2 + r^2) - (xp + yq + zr)^2$$

$$g = x^2p^2 + x^2q^2 + x^2r^2 + y^2p^2 + y^2q^2 + y^2r^2 + z^2p^2 + z^2q^2 + z^2r^2$$

$$-x^2p^2 - y^2q^2 - z^2r^2 - 2xypq - 2xzpr - 2yzqr$$

$$g = \boxed{x^2q^2 + x^2r^2 + y^2p^2 + y^2r^2 + z^2p^2 + z^2q^2 - 2xypq - 2xzpr - 2yzqr}$$

$$(B)\ x = 1\ ,\quad y = 3\ ,\quad z = 5\ ,\quad p = 2\ ,\quad q = 4\ ,\quad r = 6$$

$$f = [(1)(4) - (3)(2)]^2 + [(1)(6) - (5)(2)]^2 + [(3)(6) - (5)(4)]^2$$

$$f = (4 - 6)^2 + (6 - 10)^2 + (18 - 20)^2 = (-2)^2 + (-4)^2 + (-2)^2$$

$$f = 4 + 16 + 4 = \boxed{24}$$

$$g = (1^2 + 3^2 + 5^2)(2^2 + 4^2 + 6^2) - [(1)(2) + (3)(4) + (5)(6)]^2$$

$$g = (1 + 9 + 25)(4 + 16 + 36) - (2 + 12 + 30)^2$$

$$g = (35)(56) - (44)^2 = 1960 - 1936 = \boxed{24}\quad ✓$$

Solution to Exercise #10	
$(A)\ 2(2.4x - 1.8y) = 2(0.3)$	Multiply by 2 on both sides of the top
$4.8x - 3.6y = 0.6$	equation.
$1.2x + 3.6y = 7.8$	Add the two equations together: $3.6y$
$6x = 8.4$	cancels out.
$x = \dfrac{8.4}{6} = \boxed{1.4} = \boxed{\dfrac{7}{5}}$	Both the decimal and fractional form of the answer are given; only one form is
$2.4(1.4) - 1.8y = 0.3$	needed.
$3.36 - 1.8y = 0.3$	To solve for y, plug $x = 1.4$ into either of
$3.36 = 1.8y + 0.3$	the given equations.
$3.06 = 1.8y$	
$\dfrac{3.06}{1.8} = \boxed{1.7} = \boxed{\dfrac{17}{10}} = y$	

$$(B)\ 2.4(1.4) - 1.8(1.7) = 3.36 - 3.06 = 0.3\quad ✓$$

$$1.2x + 3.6y = 1.2(1.4) + 3.6(1.7) = 1.68 + 6.12 = 7.8\quad ✓$$

Solution to Exercise #11

(A) $u = x + \left(\dfrac{1}{2x} + \dfrac{1}{x}\right)^{-1} = x + \left(\dfrac{1}{2x} + \dfrac{2}{2x}\right)^{-1} = x + \left(\dfrac{3}{2x}\right)^{-1}$

$$u = x + \dfrac{2x}{3} = \dfrac{3x}{3} + \dfrac{2x}{3} = \boxed{\dfrac{5x}{3}}$$

(B) $t = x + \left(\dfrac{3}{5x} + \dfrac{1}{x}\right)^{-1} = x + \left(\dfrac{3}{5x} + \dfrac{5}{5x}\right)^{-1} = x + \left(\dfrac{8}{5x}\right)^{-1}$

$$t = x + \dfrac{5x}{8} = \dfrac{8x}{8} + \dfrac{5x}{8} = \boxed{\dfrac{13x}{8}}$$

(C) $s = x + \left(\dfrac{8}{13x} + \dfrac{1}{x}\right)^{-1} = x + \left(\dfrac{8}{13x} + \dfrac{13}{13x}\right)^{-1} = x + \left(\dfrac{21}{13x}\right)^{-1}$

$$s = x + \dfrac{13x}{21} = \dfrac{21x}{21} + \dfrac{13x}{21} = \boxed{\dfrac{34x}{21}}$$

(D) $r = x + \left(\dfrac{21}{34x} + \dfrac{1}{x}\right)^{-1} = x + \left(\dfrac{21}{34x} + \dfrac{34}{34x}\right)^{-1} = x + \left(\dfrac{55}{34x}\right)^{-1}$

$$r = x + \dfrac{34x}{55} = \dfrac{55x}{55} + \dfrac{34x}{55} = \boxed{\dfrac{89x}{55}}$$

(E) $q = x + \left(\dfrac{55}{89x} + \dfrac{1}{x}\right)^{-1} = x + \left(\dfrac{55}{89x} + \dfrac{89}{89x}\right)^{-1} = x + \left(\dfrac{144}{89x}\right)^{-1}$

$$q = x + \dfrac{89x}{144} = \dfrac{144x}{144} + \dfrac{89x}{144} = \boxed{\dfrac{233x}{144}}$$

(F) The numerator and denominator of the answer follow the Fibonacci sequence: 0, 1, 1, 2, 3, 5, 8, 13, 21, 34, 55, 89, 144... Each number in the Fibonacci sequence is the sum of the previous two numbers. For example, $5 + 8 = 13$ and $8 + 13 = 21$.

Solution to Exercise #12

$$\dfrac{1}{x} = \dfrac{2}{t + u}$$

$$\dfrac{1}{y} = \dfrac{2}{t - u}$$

$$\dfrac{1}{x} + \dfrac{1}{y} = \dfrac{2}{t + u} + \dfrac{2}{t - u}$$

$$\frac{1}{x} + \frac{1}{y} = \frac{2(t-u)}{(t-u)(t+u)} + \frac{2(t+u)}{(t-u)(t+u)}$$

$$\frac{1}{x} + \frac{1}{y} = \frac{2t - 2u + 2t + 2u}{t^2 + tu - tu - u^2}$$

$$\frac{1}{x} + \frac{1}{y} = \frac{4t}{t^2 - u^2}$$

$$\boxed{\frac{1}{x} + \frac{1}{y}} = \frac{4t}{4tz} = \boxed{\frac{1}{z}}$$

Solution to Exercise #13

(A) $x_1 y_1 = kz_1$, $x_2 y_2 = kz_2$

$$\frac{x_1 y_1}{k} = z_1 \quad , \quad \frac{x_2 y_2}{k} = z_2$$

$$z_1 = z_2$$

$$\frac{x_1 y_1}{k} = \frac{x_2 y_2}{k}$$

$$\boxed{x_1 y_1 = x_2 y_2}$$

(B) $x_1 y_1 = kz_1$, $x_2 y_2 = kz_2$

$$y_1 = \frac{kz_1}{x_1} \quad , \quad y_2 = \frac{kz_2}{x_2}$$

$$y_1 = y_2$$

$$\frac{kz_1}{x_1} = \frac{kz_2}{x_2}$$

$$\frac{kz_1}{x_1} = \frac{kz_2}{x_2} \quad \text{or} \quad \boxed{\frac{x_1}{z_1} = \frac{x_2}{z_2}}$$

(C) $x_1 y_1 = kz_1$, $x_2 y_2 = kz_2$

$$x_1 = \frac{kz_1}{y_1} \quad , \quad z_2 = \frac{kz_2}{y_2}$$

$$x_1 = x_2$$

$$\frac{kz_1}{y_1} = \frac{kz_2}{y_2}$$

$$\frac{z_1}{y_1} = \frac{z_2}{y_2} \quad \text{or} \quad \boxed{\frac{y_1}{z_1} = \frac{y_2}{z_2}}$$

(D) $x_1 y_1 = kz_1$, $x_2 y_2 = kz_2$

$$\frac{x_1 y_1}{z_1} = k \quad , \quad \frac{x_2 y_2}{z_2} = k$$

$$\boxed{\frac{x_1 y_1}{z_1} = \frac{x_2 y_2}{z_2}}$$

Notes: These formulas are similar to Boyle's law ($P_1 V_1 = P_2 V_2$ or $PV = $ const. if T is constant), Charles's law ($\frac{V_1}{T_1} = \frac{V_2}{T_2}$ or $\frac{V}{T} = $ const. if P is constant), Gay-Lussac's law ($\frac{P_1}{T_1} = \frac{P_2}{T_2}$ or $\frac{P}{T} = $ const. if V is constant), and the ideal gas law ($\frac{P_1 V_1}{T_1} = \frac{P_2 V_2}{T_2}$ or $\frac{PV}{T} = nR = $ const.). Note, for example, that $\frac{z_1}{x_1} = \frac{z_2}{x_2}$ is equivalent to $\frac{x_1}{z_1} = \frac{x_2}{z_2}$; to see this, take the reciprocal of both sides of the equation.

Solution to Exercise #14

$$(A) \; (3x^2 + 4x - 2)^3 - (2x^2 - 3x + 4)^3$$

$$= (3x^2 + 4x - 2)(3x^2 + 4x - 2)^2 - (2x^2 - 3x + 4)(2x^2 - 3x + 4)^2$$

$$= (3x^2 + 4x - 2)(9x^4 + 16x^2 + 4 + 24x^3 - 12x^2 - 16x)$$

$$-(2x^2 - 3x + 4)(4x^4 + 9x^2 + 16 - 12x^3 + 16x^2 - 24x)$$

$$= (3x^2 + 4x - 2)(9x^4 + 24x^3 + 4x^2 - 16x + 4)$$

$$-(2x^2 - 3x + 4)(4x^4 - 12x^3 + 25x^2 - 24x + 16)$$

$$= 27x^6 + 72x^5 + 12x^4 - 48x^3 + 12x^2$$

$$+ \, 36x^5 + 96x^4 + 16x^3 - 64x^2 + 16x$$

$$-18x^4 - 48x^3 - 8x^2 + 32x - 8$$

$$- \, 8x^6 + 24x^5 - 50x^4 + 48x^3 - 32x^2$$

$$+ \, 12x^5 - 36x^4 + 75x^3 - 72x^2 + 48x$$

$$- \, 16x^4 + 48x^3 - 100x^2 + 96x - 64$$

$$= (27 - 8)x^6 + (72 + 36 + 24 + 12)x^5 + (12 + 96 - 18 - 50 - 36 - 16)x^4$$

$$+(-48 + 16 - 48 + 48 + 75 + 48)x^3 + (12 - 64 - 8 - 32 - 72 - 100)x^2$$

$$+(16 + 32 + 48 + 96)x - 8 - 64$$

$$= \boxed{19x^6 + 144x^5 - 12x^4 + 91x^3 - 264x^2 + 192x - 72}$$

$$(B) \; 19x^6 + 144x^5 - 12x^4 + 91x^3 - 264x^2 + 192x - 72$$

$$= 19\left(\frac{1}{2}\right)^6 + 144\left(\frac{1}{2}\right)^5 - 12\left(\frac{1}{2}\right)^4 + 91\left(\frac{1}{2}\right)^3 - 264\left(\frac{1}{2}\right)^2 + 192\left(\frac{1}{2}\right) - 72$$

$$= \frac{19}{64} + \frac{144}{32} - \frac{12}{16} + \frac{91}{8} - \frac{264}{4} + \frac{192}{2} - 72$$

$$= \frac{19}{64} + \frac{288}{64} - \frac{48}{64} + \frac{728}{64} - 61 + 96 - 72$$

$$= \frac{987}{64} - 37 = 15 + \frac{27}{64} - 42 = \boxed{\frac{27}{64} - 27}$$

$$= \frac{27}{64} - \frac{1728}{64} = \boxed{-\frac{1701}{64}}$$

$$(3x^2 + 4x - 2)^3 - (2x^2 - 3x + 4)^3$$

$$= \left[3\left(\frac{1}{2}\right)^2 + 4\left(\frac{1}{2}\right) - 2\right]^3 - \left[2\left(\frac{1}{2}\right)^2 - 3\left(\frac{1}{2}\right) + 4\right]^3$$

$$= \left[3\left(\frac{1}{4}\right) + 2 - 2\right]^3 - \left[2\left(\frac{1}{4}\right) - \frac{3}{2} + 4\right]^3$$

$$= \left(\frac{3}{4} + 0\right)^3 - \left(\frac{1}{2} - \frac{3}{2} + \frac{8}{2}\right)^3$$

$$= \frac{3^3}{4^3} - \left(\frac{6}{2}\right)^3 = \frac{27}{64} - 3^3 = \boxed{\frac{27}{64} - 27} \quad \checkmark$$

$$= \frac{27}{64} - \frac{1728}{64} = \boxed{-\frac{1701}{64}} \quad \checkmark$$

Solution to Exercise #15

(A) $z = \dfrac{x}{y}$

$yz = x$

$y = \dfrac{x}{z}$

$\boxed{w} = \dfrac{xy}{2} = \dfrac{x}{2}\left(\dfrac{x}{z}\right) = \boxed{\dfrac{x^2}{2z}}$

(B) $z = \dfrac{x}{y}$

$yz = x$

$\boxed{w} = \dfrac{xy}{2} = \dfrac{(yz)y}{2} = \boxed{\dfrac{y^2 z}{2}}$

(C) $w < 9$, $z = 2$

$w = \dfrac{x^2}{2z}$

$x^2 = 2wz$

$x^2 < 2(9)(2)$

$x^2 < 36$

$x < \sqrt{36}$ or $x > -\sqrt{36}$

$x < 6$ or $x > -6$

$\boxed{-6 < x < 6}$

See the note on the following page.

$w = \dfrac{y^2 z}{2}$

$y^2 = \dfrac{2w}{z}$

$y^2 < \dfrac{2(9)}{2}$

$y^2 < 9$

$y < \sqrt{9}$ or $y > -\sqrt{9}$

$y < 3$ or $y > -3$

$\boxed{-3 < y < 3}$

Note: $x < 6$ or $x > -6$ satisfies $x^2 < 36$. For example, if $x = -5$, this satisfies $x > -6$ and also satisfies $x^2 < 36$ because $(-5)^2 = 25 < 36$. In contrast, $x = -7$ does NOT satisfy $x > -6$ and also does NOT satisfy $x^2 < 36$ because $(-7)^2 = 49 > 36$. A helpful way to check the answer to an algebra problem with an inequality is to plug in values that slightly satisfy and slightly do not satisfy the inequality for the answer into the original inequality.

Note: These formulas are similar to the equations for capacitance: $C = \frac{Q}{V}$ and

$$U = \frac{QV}{2} = \frac{CV^2}{2} = \frac{Q^2}{2C}.$$

Solution to Exercise #16

(A) $2z = 3$

$$z = \boxed{\frac{3}{2}} = \boxed{1.5}$$

$$2y + 3z = 4$$
$$2y + 3(1.5) = 4$$
$$2y + 4.5 = 4$$
$$2y = -0.5$$
$$y = -\frac{0.5}{2} = \boxed{-0.25} = \boxed{-\frac{1}{4}}$$

$$2x + 3y + 4z = 5$$
$$2x + 3(-0.25) + 4(1.5) = 5$$
$$2x - 0.75 + 6 = 5$$
$$2x + 5.25 = 5$$
$$2x = -0.25$$
$$x = -\frac{0.25}{2} = \boxed{-0.125} = \boxed{-\frac{1}{8}}$$

$$2w + 3x + 4y + 5z = 6$$
$$2w + 3(-0.125) + 4(-0.25) + 5(1.5) = 6$$
$$2w - 0.375 - 1 + 7.5 = 6$$

(B) $2w + 3x + 4y + 5z = 6$
$$2(-0.0625) + 3(-0.125) + 4(-0.25) + 5(1.5) = 6$$
$$-0.125 - 0.375 - 1 + 7.5 = 6 \quad \checkmark$$

$$2x + 3y + 4z = 5$$
$$2(-0.125) + 3(-0.25) + 4(1.5) = 5$$
$$-0.25 - 0.75 + 6 = 5 \quad \checkmark$$

$$2y + 3z = 4$$
$$2(-0.25) + 3(1.5) = 4$$
$$-0.5 + 4.5 = 4 \quad \checkmark$$

$$2z = 3$$
$$2(1.5) = 3 \quad \checkmark$$

(C) $y = -\frac{1}{4}$, $x = -\frac{1}{8}$, $w = -\frac{1}{16}$ form a pattern (excluding z): divide by two. For example, $-\frac{1}{4} \div 2 = -\frac{1}{4} \times \frac{1}{2} = -\frac{1}{8}$. If the equation $2u + 3w + 4x + 5y + 6z = 7$ is added to the system (which continues the pattern of equations), the answer for the new variable would be $u = -\frac{1}{32}$.

$$2w + 6.125 = 6$$

$$2w = -0.125$$

$$w = -\frac{0.125}{2} = \boxed{-0.0625} = \boxed{-\frac{1}{16}}$$

Solution to Exercise #17	
(A) $\dfrac{\sqrt{x}}{\sqrt{x} + \sqrt{3}} + \dfrac{\sqrt{3}}{\sqrt{x} - \sqrt{3}}$	Make a common denominator.
$= \dfrac{\sqrt{x}}{\sqrt{x} + \sqrt{3}}\left(\dfrac{\sqrt{x} - \sqrt{3}}{\sqrt{x} - \sqrt{3}}\right) + \dfrac{\sqrt{3}}{\sqrt{x} - \sqrt{3}}\left(\dfrac{\sqrt{x} + \sqrt{3}}{\sqrt{x} + \sqrt{3}}\right)$	$(a + b)(a - b) = a^2 - ab + ab - b^2$ $= a^2 - b^2$
$= \dfrac{\sqrt{x}\sqrt{x} - \sqrt{x}\sqrt{3} + \sqrt{3}\sqrt{x} + \sqrt{3}\sqrt{3}}{\sqrt{x}\sqrt{x} - \sqrt{x}\sqrt{3} + \sqrt{3}\sqrt{x} - \sqrt{3}\sqrt{3}}$	Note that $\sqrt{x}\sqrt{x} = x$ and $\sqrt{3}\sqrt{3} = 3$.
$= \boxed{\dfrac{x + 3}{x - 3}}$	
(B) $\dfrac{x + 3}{x - 3} = \dfrac{5}{3}$	Cross multiply:
$3(x + 3) = 5(x - 3)$	$\dfrac{a}{b} = \dfrac{c}{d}$
$3x + 9 = 5x - 15$	$ad = bc$
$9 + 15 = 5x - 3x$	
$24 = 2x$	
$\dfrac{24}{2} = \boxed{12} = x$	
(C) $\dfrac{\sqrt{12}}{\sqrt{12} + \sqrt{3}} + \dfrac{\sqrt{3}}{\sqrt{12} - \sqrt{3}}$	$\sqrt{12} = \sqrt{(4)(3)} = \sqrt{4}\sqrt{3} = 2\sqrt{3}$
$= \dfrac{2\sqrt{3}}{2\sqrt{3} + \sqrt{3}} + \dfrac{\sqrt{3}}{2\sqrt{3} - \sqrt{3}}$	$2\sqrt{3} + \sqrt{3} = (2 + 1)\sqrt{3} = 3\sqrt{3}$ $2\sqrt{3} - \sqrt{3} = (2 - 1)\sqrt{3} = \sqrt{3}$
$= \dfrac{2\sqrt{3}}{3\sqrt{3}} + \dfrac{\sqrt{3}}{\sqrt{3}} = \dfrac{2}{3} + 1 = \dfrac{2}{3} + \dfrac{3}{3} = \dfrac{5}{3}$ ✔	
(D) $\sqrt{x} - \sqrt{3} = 0$	If $\sqrt{x} - \sqrt{3}$ is zero, this would cause
$\sqrt{x} = \sqrt{3}$	division by zero in the given
$x = \boxed{3}$	expression.

Solution to Exercise #18

(A) $16x - 3 = 8\sqrt{x}$

$16x - 8\sqrt{x} - 3 = 0$

$16y^2 - 8y - 3 = 0$

$(4y - 3)(4y + 1) = 0$

$4y - 3 = 0$ or $4y + 1 = 0$

$4y = 3$ or $4y = -1$

$y = \dfrac{3}{4}$ or $y = -\dfrac{1}{4}$

$x = y^2$

$x = \left(\dfrac{3}{4}\right)^2$ or $x = \left(-\dfrac{1}{4}\right)^2$

$x = \boxed{\dfrac{9}{16}} = \boxed{0.5625}$ only

(B) $16\left(\dfrac{9}{16}\right) - 3 = 9 - 3 = 6$

$8\sqrt{\dfrac{9}{16}} = 8\left(\dfrac{3}{4}\right) = 2(3) = 6$ ✓

One way to solve this problem is to let $y = \sqrt{x}$. This gives a quadratic equation for y. The quadratic equation can either be solved by factoring (as shown to the left) or by using the quadratic formula. After solving for y, plug each answer for y into the equation $y = \sqrt{x}$, which is equivalent to $y^2 = x$, to solve for x. Another way to solve this problem is to square both sides of the given equation. Use the "foil" method to square the left-hand side:

$$(16x - 3)^2 = 256x^2 - 96x + 9$$

The answer $x = \left(-\dfrac{1}{4}\right)^2 = \dfrac{1}{16}$ would only satisfy the given equation if $\sqrt{x} = -\dfrac{1}{4}$, but the problem states that $\sqrt{x} > 0$. Therefore, $x = \left(-\dfrac{1}{4}\right)^2$ is NOT a valid answer for this problem.

Solution to Exercise #19

(A) $z = \dfrac{w - u}{y}$

$yz = w - u$

$\boxed{yz + u = w}$

(B) $x = \left(\dfrac{u + w}{2}\right) y = \left(\dfrac{u + yz + u}{2}\right) y$

(C) $z = \dfrac{w - u}{y}$

$yz = w - u$

$y = \dfrac{w - u}{z}$

$x = \left(\dfrac{u + w}{2}\right) y = \left(\dfrac{u + w}{2}\right)\left(\dfrac{w - u}{z}\right)$

$x = \dfrac{w^2 - u^2}{2z}$

$$x = \left(\frac{yz + 2u}{2}\right)y = \left(\frac{yz}{2} + u\right)y$$

$$= \frac{y^2 z}{2} + uy$$

$$2xz = w^2 - u^2$$

$$u^2 + 2xz = w^2$$

$$w = \pm\sqrt{u^2 + 2xz}$$

Note: Unlike the previous problem, the instructions for this problem did not restrict the possible answers. When taking the square root of both sides of $u^2 + 2xz = w^2$, it is necessary to consider both possible signs of the square root in order to find all of the solutions.

Solution to Exercise #20	
(A) $x^2 - 3x + 7x - 21 = x^2 - 10x + 25$ $x^2 + 4x - 21 = x^2 - 10x + 25$ $4x + 10x = 25 + 21$ $14x = 46$ $x = \dfrac{46}{14} = \boxed{\dfrac{23}{7}}$	Apply the "foil" method: $(a + b)(c - d) = ac - ad + bc - bd$ Subtract x^2 from both sides, add $10x$ to both sides, and add 21 to both sides. Note that x^2 cancels out.
(B) $\left(\dfrac{23}{7} + 7\right)\left(\dfrac{23}{7} - 3\right) = \left(\dfrac{23}{7} - 5\right)^2$ $\left(\dfrac{23}{7} + \dfrac{49}{7}\right)\left(\dfrac{23}{7} - \dfrac{21}{7}\right) = \left(\dfrac{23}{7} - \dfrac{35}{7}\right)^2$	Make common denominators.
$\left(\dfrac{72}{7}\right)\left(\dfrac{2}{7}\right) = \left(-\dfrac{12}{7}\right)^2$ $\dfrac{144}{49} = \dfrac{144}{49}$ ✔	The square of a negative number is positive.

Solution to Exercise #21

(A) $\dfrac{2x}{x-3} - \dfrac{4-x}{x}$

Make a common denominator of $(x-3)x$.

$= \dfrac{2x}{x-3}\left(\dfrac{x}{x}\right) - \dfrac{4-x}{x}\left(\dfrac{x-3}{x-3}\right)$

$= \dfrac{2x(x) - (4-x)(x-3)}{(x-3)(x)}$

Apply the "foil" method:
$$(a-b)(c-d) = ac - ad - bc + bd$$

$= \dfrac{2x^2 - (4x - 12 - x^2 + 3x)}{x^2 - 3x}$

Combine like terms: $4x + 3x = 7x$.

$= \dfrac{2x^2 - (-x^2 + 7x - 12)}{x^2 - 3x}$

Distribute the minus sign:
$$-(-x^2 + 7x - 12)$$

$= \dfrac{2x^2 + x^2 - 7x + 12}{x^2 - 3x}$

$$= -(-x^2) - (7x) - (-12)$$
$$x^2 - 7x + 12$$

$= \dfrac{3x^2 - 7x + 12}{x^2 - 3x} = \dfrac{P}{Q}$

$\boxed{P = 3x^2 - 7x + 12}$

$\boxed{Q = x^2 - 3x}$

(B) $P = 3(9)^2 - 7(9) + 12$

$P = 3(81) - 63 + 12$

$P = 243 - 51 = 192$

$Q = (9)^2 - 3(9)$

$Q = 81 - 27 = 54$

$\dfrac{P}{Q} = \dfrac{192}{54} = \boxed{\dfrac{32}{9}} \quad \checkmark$

Divide 192 and 54 each by 6 to reduce the fraction.

$\dfrac{2x}{x-3} - \dfrac{4-x}{x} = \dfrac{2(9)}{9-3} - \dfrac{4-9}{9}$

$= \dfrac{18}{6} - \dfrac{-5}{9} = \dfrac{54}{18} + \dfrac{10}{18} = \dfrac{64}{18} = \boxed{\dfrac{32}{9}} \quad \checkmark$

The two minus signs make a plus sign.

(C) $x = 0$ and $x = 3$ would cause division by zero.

Solution to Exercise #22

(A) $\dfrac{x-6}{y+3} = \dfrac{3}{4}$	Cross multiply:
	$\dfrac{a}{b} = \dfrac{c}{d}$
$4(x-6) = 3(y+3)$	
$4x - 24 = 3y + 9$	$ad = bc$
$4x - 3y = 33$	Equation 1.
$\dfrac{y+7}{x+9} = \dfrac{2}{3}$	Cross multiply again.
$3(y+7) = 2(x+9)$	
$3y + 21 = 2x + 18$	
$-2x + 3y = -3$	Equation 2.
$4x - 3y = 33$	Equation 1 from above.
$-2x + 4x = -3 + 33$	Add Equation 1 and Equation 2 together:
$2x = 30$	$3y$ cancels out.
$x = \boxed{15}$	
$-2x + 3y = -3$	Equation 2 from above.
$-2(15) + 3y = -3$	Plug $x = 15$ into Equation 2.
$-30 + 3y = -3$	
$3y = 27$	
$y = \boxed{9}$	
(B) $\dfrac{x-6}{y+3} = \dfrac{15-6}{9+3} = \dfrac{9}{12} = \dfrac{3}{4}$ ✓	
$\dfrac{y+7}{x+9} = \dfrac{9+7}{15+9} = \dfrac{16}{24} = \dfrac{2}{3}$ ✓	Plug $x = 15$ and $y = 9$ into each of the given equations.

Solution to Exercise #23

(A) $u = \dfrac{z^2}{x}$	Multiply both sides by x.
$ux = z^2$	
$5y = 10x + \dfrac{z^2}{4}$	Replace z^2 in this equation with ux.

$$5y = 10x + \frac{ux}{4}$$

$$w + 5 = \frac{u}{2}$$

Multiply both sides by 2.

$$u = 2w + 10$$

This is equivalent to $2w + 10 = u$.

$$w > 0$$

Since $w > 0$, it follows that $2w > 0$ and

$$u > 2(0) + 10$$

that $2w + 10 > 10$, which shows that

$$u > 10$$

$u > 10$.

$$5y = 10x + \frac{ux}{4}$$

See the top of this page.

$$5y > 10x + \frac{10x}{4}$$

Plug $u > 10$ into the previous equation.

$$5y > \frac{40x}{4} + \frac{10x}{4}$$

Make a common denominator.

$$5y > \frac{50x}{4}$$

$\frac{50}{4}$ reduces to $\frac{25}{2}$.

$$5y > \frac{25x}{2}$$

Divide both sides by 5:

$$\boxed{y > \frac{5x}{2}} \quad \text{or} \quad \boxed{y > 2.5x}$$

$$\frac{25x}{2} \div 5 = \frac{25x}{2} \times \frac{1}{5} = \frac{25x}{10} = \frac{5x}{2} = 2.5x$$

(B) $D = 2x$

Diameter is twice the radius.

$$\frac{D}{2} = x$$

Half the diameter equals the radius.

$$y > \frac{5}{2}\left(\frac{D}{2}\right)$$

Replace x with $\frac{D}{2}$ in $y > \frac{5x}{2}$.

$$\boxed{y > \frac{5D}{4}} \quad \text{or} \quad \boxed{y > 1.25D}$$

Check that the results make sense: Since D is larger than x, a smaller number (1.25) times D equates to a larger number (2.5) times x. That is, $1.25D = 2.5x$ because $x = \frac{D}{2}$.

(C) $5y = 10x + \frac{z^2}{4}$

$$5y - \frac{z^2}{4} = 10x$$

Divide both sides by 4:

$$\frac{5y}{4} - \frac{z^2}{16} = \frac{5x}{2} < y$$

$$\frac{1}{4} \div 4 = \frac{1}{4} \times \frac{1}{4} = \frac{1}{16}$$

$$10 \div 4 = 10 \times \frac{1}{4} = \frac{10}{4} = \frac{5}{2}$$

$$\frac{5y}{4} - \frac{z^2}{16} < y$$

$$\frac{5y}{4} - y < \frac{z^2}{16}$$

$$\frac{5y}{4} - \frac{4y}{4} < \frac{z^2}{16}$$

$$\frac{y}{4} < \frac{z^2}{16}$$

$$4y < z^2$$

$$\boxed{0 \le 2\sqrt{y} < z}$$

$\frac{5x}{2} < y$ is equivalent to $y > \frac{5x}{2}$.

Since $\frac{5y}{4} - \frac{z^2}{16} = \frac{5x}{2}$ and $\frac{5x}{2} < y$, it follows that $\frac{5y}{4} - \frac{z^2}{16} < y$. (Simply replace $\frac{5x}{2}$ with $\frac{5y}{4} - \frac{z^2}{16}$ in $\frac{5x}{2} < y$.)

Multiply both sides by 16:

$$\frac{16}{4} = 4$$

Square root both sides of the equation:

$$\sqrt{4y} = \sqrt{4}\sqrt{y} = 2\sqrt{y}$$

Note: The math for Part A is similar to a classic roller coaster problem in physics. What minimum height relative to the bottom of a loop is needed if a roller coaster begins from rest (neglecting resistive forces like air resistance)? In physics, the equations look like $mgh_i = mg2R + \frac{1}{2}mv^2$, $N + mg = ma_c$, $a_c = \frac{v^2}{R}$, and $N > 0$.

Solution to Exercise #24

(A) $z = \left[\dfrac{(64x^6y^9)^{2/3}}{(2x^2y)^3\sqrt{324x^4y^{10}}}\right]^{-\frac{5}{2}}$

$z = \left[\dfrac{64^{2/3}(x^6)^{2/3}(y^9)^{2/3}}{2^3(x^2)^3y^3\sqrt{324}\sqrt{x^4}\sqrt{y^{10}}}\right]^{-\frac{5}{2}}$

$z = \left[\dfrac{\left(\sqrt[3]{64}\right)^2 x^4 y^6}{8x^6y^3 18x^2y^5}\right]^{-\frac{5}{2}}$

$z = \left(\dfrac{4^2x^4y^6}{144x^{6+2}y^{3+5}}\right)^{-\frac{5}{2}} = \left(\dfrac{16x^4y^6}{144x^8y^8}\right)^{-\frac{5}{2}}$

$z = \left(\dfrac{x^{4-8}y^{6-8}}{9}\right)^{-\frac{5}{2}} = \left(\dfrac{x^{-4}y^{-2}}{9}\right)^{-\frac{5}{2}}$

Recall the rules $(a^b)^c = a^{bc}$, $(pq)^r = p^r q^r$, and $\sqrt{uw} = \sqrt{u}\sqrt{w}$.

$6\left(\frac{2}{3}\right) = \frac{12}{3} = 4$ and $9\left(\frac{2}{3}\right) = \frac{18}{3} = 6$.

$64^{2/3} = \left(\sqrt[3]{64}\right)^2 = 4^2 = 16$.

$\sqrt{t^n} = t^{n/2}$ because $t^{n/2}t^{n/2} = t^n$.

Recall the rule $u^m u^n = u^{m+n}$.

Recall the rule $\dfrac{w^m}{w^n} = w^{m-n}$.

$$z = \left(\frac{1}{9x^4y^2}\right)^{-\frac{5}{2}} = (9x^4y^2)^{\frac{5}{2}}$$

$$z = \left(\sqrt{9x^4y^2}\right)^5 = (3x^2y)^5$$

$$z = 3^5x^{10}y^5 = 243x^{10}y^5$$

$$z = ax^by^c$$

$$a = \boxed{243} \quad , \quad b = \boxed{10} \quad , \quad c = \boxed{5}$$

$$(B) \; z = 243\left(\frac{1}{2}\right)^{10}(4)^5 = \frac{(243)(4^5)}{2^{10}}$$

$$z = \frac{243(1024)}{1024} = \boxed{243}$$

A negative exponent involves a reciprocal: $\left(\frac{p}{q}\right)^{-1} = \frac{q}{p}$ and $\left(\frac{p}{q}\right)^{-k} = \left(\frac{q}{p}\right)^k = \frac{q^k}{p^k}.$

Recall the rule $(r^m)^n = r^{mn}$.

$2^{10} = 1024$ and $4^5 = 1024$.

$$z = \left\{\frac{\left[64\left(\frac{1}{2}\right)^6(4)^9\right]^{2/3}}{\left[2\left(\frac{1}{2}\right)^2(4)\right]^3\sqrt{324\left(\frac{1}{2}\right)^4(4)^{10}}}\right\}^{-\frac{5}{2}}$$

$2^6 = 64$, $4^9 = 262{,}144$, and $4^{10} = 1{,}048{,}576$.

$$z = \left\{\frac{\left[64\left(\frac{1}{64}\right)(262{,}144)\right]^{2/3}}{\left[2\left(\frac{1}{4}\right)(4)\right]^3\sqrt{324\left(\frac{1}{16}\right)(1{,}048{,}576)}}\right\}^{-\frac{5}{2}}$$

$$z = \left(\frac{262{,}144^{2/3}}{2^3\sqrt{21{,}233{,}664}}\right)^{-\frac{5}{2}} = \left[\frac{\left(\sqrt[3]{262{,}144}\right)^2}{8(4608)}\right]^{-\frac{5}{2}}$$

$t^{2/3} = \left(\sqrt[3]{t}\right)^2$

$$z = \left[\frac{(64)^2}{36{,}864}\right]^{-\frac{5}{2}} = \left(\frac{4096}{36{,}864}\right)^{-\frac{5}{2}} = \left(\frac{1}{9}\right)^{-\frac{5}{2}}$$

$\left(\frac{p}{q}\right)^{-k} = \left(\frac{q}{p}\right)^k = \frac{q^k}{p^k}$

$$z = (9)^{\frac{5}{2}} = \left(\sqrt{9}\right)^5 = (3)^5 = \boxed{243} \quad \checkmark$$

$u^{5/2} = \left(\sqrt{u}\right)^5$

Solution to Exercise #25	
(A) $7x + 3y - 11 = z$	(B) $7(0.44) + 3(-6.48) - (-27.36)$
$-4x + 9y - 3(7x + 3y - 11) = 22$	$= 3.08 - 19.44 + 27.36 = 11$ ✓
$-4x + 9y - 21x - 9y + 33 = 22$	$-4(0.44) + 9(-6.48) - 3(-27.36)$
$-25x = -11$	$= -1.76 - 58.32 + 82.08 = 22$ ✓

$$x = \boxed{\frac{11}{25}} = \boxed{0.44}$$

$$3x - 5y + 2(7x + 3y - 11) = -21$$

$$3x - 5y + 14x + 6y - 22 = -21$$

$$17x + y = 1$$

$$17\left(\frac{11}{25}\right) + y = 1$$

$$\frac{187}{25} + y = 1$$

$$y = \frac{25}{25} - \frac{187}{25}$$

$$y = \boxed{-\frac{162}{25}} = \boxed{-6.48}$$

$$z = 7x + 3y - 11$$

$$z = 7\left(\frac{11}{25}\right) + 3\left(-\frac{162}{25}\right) - 11$$

$$z = \frac{77}{25} - \frac{486}{25} - \frac{275}{25}$$

$$z = \boxed{-\frac{684}{25}} = \boxed{-27.36}$$

$$3(0.44) - 5(-6.48) + 2(-27.36)$$

$$= 1.32 + 32.4 - 54.72 = -21 \quad \checkmark$$

Solution to Exercise #26

(A) $\dfrac{2}{25 - 4x} = \dfrac{x}{3}$

$$2(3) = x(25 - 4x)$$

$$6 = 25x - 4x^2$$

$$4x^2 - 25x + 6 = 0$$

$$(4x - 1)(x - 6) = 0$$

$$4x - 1 = 0 \quad \text{or} \quad x - 6 = 0$$

$$4x = 1 \quad \text{or} \quad x = 6$$

$$x = \boxed{\frac{1}{4}} = \boxed{0.25} \quad \text{or} \quad x = \boxed{6}$$

Cross multiply:

$$\frac{a}{b} = \frac{c}{d}$$

$$ad = bc$$

Put the quadratic equation in standard form. Either factor the quadratic (as shown to the left) or use the quadratic formula.

There are two possible answers for x:

$\frac{1}{4}$ or 6.

(B) $\dfrac{2}{25-4x} = \dfrac{2}{25-4\left(\frac{1}{4}\right)}$

$= \dfrac{2}{25-1} = \dfrac{2}{24} = \dfrac{1}{12}$

$\dfrac{x}{3} = \dfrac{1/4}{3} = \dfrac{1}{4} \div 3 = \dfrac{1}{4} \times \dfrac{1}{3} = \dfrac{1}{12}$ ✓

$\dfrac{2}{25-4x} = \dfrac{2}{25-4(6)}$

$= \dfrac{2}{25-24} = \dfrac{2}{1} = 2$

$\dfrac{x}{3} = \dfrac{6}{3} = 2$ ✓

Plug $x = \frac{1}{4}$ into the given equation.

To divide a fraction by a number, multiply the fraction by the reciprocal of the number: $\dfrac{a}{b} \div c = \dfrac{a}{b} \times \dfrac{1}{c} = \dfrac{a}{bc}$.

Plug $x = 6$ into the given equation.

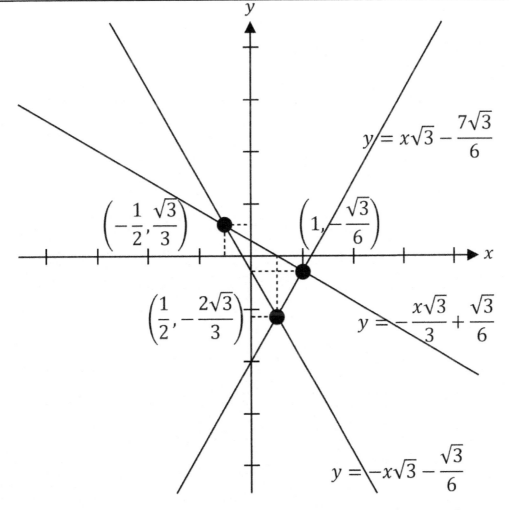

$y = x\sqrt{3} - \dfrac{7\sqrt{3}}{6}$

$\left(-\dfrac{1}{2}, \dfrac{\sqrt{3}}{3}\right)$

$\left(1, -\dfrac{\sqrt{3}}{6}\right)$

$\left(\dfrac{1}{2}, -\dfrac{2\sqrt{3}}{3}\right)$

$y = -\dfrac{x\sqrt{3}}{3} + \dfrac{\sqrt{3}}{6}$

$y = -x\sqrt{3} - \dfrac{\sqrt{3}}{6}$

Solution to Exercise #27

(A) $x\sqrt{3} - \dfrac{7\sqrt{3}}{6} = -x\sqrt{3} - \dfrac{\sqrt{3}}{6}$

| Set two of the equations equal.

$x - \dfrac{7}{6} = -x - \dfrac{1}{6}$

| Divide both sides by $\sqrt{3}$.

$2x = \dfrac{6}{6} = 1$

$$-\dfrac{1}{6} + \dfrac{7}{6} = \dfrac{6}{6} = 1$$

Divide both sides by 2.

$x = \dfrac{1}{2}$

Plug $x = \dfrac{1}{2}$ into $y = x\sqrt{3} - \dfrac{7\sqrt{3}}{6}$.

$y = \dfrac{\sqrt{3}}{2} - \dfrac{7\sqrt{3}}{6} = \dfrac{3\sqrt{3}}{6} - \dfrac{7\sqrt{3}}{6}$

Make a common denominator.

$y = -\dfrac{4\sqrt{3}}{6} = -\dfrac{2\sqrt{3}}{3}$

$(x, y) = \boxed{\left(\dfrac{1}{2}, -\dfrac{2\sqrt{3}}{3}\right)}$

$x\sqrt{3} - \dfrac{7\sqrt{3}}{6} = -\dfrac{x\sqrt{3}}{3} + \dfrac{\sqrt{3}}{6}$

Set two of the equations equal.

$x - \dfrac{7}{6} = -\dfrac{x}{3} + \dfrac{1}{6}$

Divide both sides by $\sqrt{3}$.

$\dfrac{4x}{3} = \dfrac{8}{6} = \dfrac{4}{3}$

$$\dfrac{1}{6} + \dfrac{7}{6} = \dfrac{8}{6} = \dfrac{4}{3}$$

$x = 1$

Multiply both sides by $\dfrac{3}{4}$.

$y = 1\sqrt{3} - \dfrac{7\sqrt{3}}{6} = \dfrac{6\sqrt{3}}{6} - \dfrac{7\sqrt{3}}{6}$

Plug $x = 1$ into $y = x\sqrt{3} - \dfrac{7\sqrt{3}}{6}$.

Make a common denominator.

$y = -\dfrac{\sqrt{3}}{6}$

$(x, y) = \boxed{\left(1, -\dfrac{\sqrt{3}}{6}\right)}$

$-x\sqrt{3} - \dfrac{\sqrt{3}}{6} = -\dfrac{x\sqrt{3}}{3} + \dfrac{\sqrt{3}}{6}$

Set two of the equations equal.

$-x - \dfrac{1}{6} = -\dfrac{x}{3} + \dfrac{1}{6}$

Divide both sides by $\sqrt{3}$.

$$-\frac{2x}{3} = \frac{2}{6} = \frac{1}{3}$$

$$x = -\frac{1}{2}$$

$$y = \frac{1\sqrt{3}}{6} + \frac{\sqrt{3}}{6} = \frac{2\sqrt{3}}{6} = \frac{\sqrt{3}}{3}$$

$$(x, y) = \boxed{\left(-\frac{1}{2}, \frac{\sqrt{3}}{3}\right)}$$

(B) $-\frac{2\sqrt{3}}{3} = \frac{1\sqrt{3}}{2} - \frac{7\sqrt{3}}{6} = \frac{3\sqrt{3}}{6} - \frac{7\sqrt{3}}{6}$

$$= -\frac{4\sqrt{3}}{6} = -\frac{2\sqrt{3}}{3} \quad \checkmark$$

$-\frac{2\sqrt{3}}{3} = -\frac{1\sqrt{3}}{2} - \frac{\sqrt{3}}{6} = -\frac{3\sqrt{3}}{6} - \frac{\sqrt{3}}{6}$

$$= -\frac{4\sqrt{3}}{6} = -\frac{2\sqrt{3}}{3} \quad \checkmark$$

$$-\frac{\sqrt{3}}{6} = 1\sqrt{3} - \frac{7\sqrt{3}}{6}$$

$$= \frac{6\sqrt{3}}{6} - \frac{7\sqrt{3}}{6} = -\frac{\sqrt{3}}{6} \quad \checkmark$$

$$-\frac{\sqrt{3}}{6} = -\frac{1\sqrt{3}}{3} + \frac{\sqrt{3}}{6}$$

$$= -\frac{2\sqrt{3}}{6} + \frac{\sqrt{3}}{6} = -\frac{\sqrt{3}}{6} \quad \checkmark$$

$\frac{\sqrt{3}}{3} = -\frac{(-1\sqrt{3})}{2} - \frac{\sqrt{3}}{6} = \frac{3\sqrt{3}}{6} - \frac{\sqrt{3}}{6}$

$$= \frac{2\sqrt{3}}{6} = \frac{\sqrt{3}}{3} \quad \checkmark$$

$\frac{\sqrt{3}}{3} = -\frac{(-1\sqrt{3})}{2(3)} + \frac{\sqrt{3}}{6} = \frac{\sqrt{3}}{6} + \frac{\sqrt{3}}{6}$

$$= \frac{2\sqrt{3}}{6} = \frac{\sqrt{3}}{3} \quad \checkmark$$

$$-\frac{1}{6} + \frac{1}{6} = \frac{2}{6} = \frac{1}{3}$$

Multiply both sides by $-\frac{3}{2}$.

Plug $x = -\frac{1}{2}$ into $y = -\frac{x\sqrt{3}}{3} + \frac{\sqrt{3}}{6}$.

Plug $\left(\frac{1}{2}, -\frac{2\sqrt{3}}{3}\right)$ into $y = x\sqrt{3} - \frac{7\sqrt{3}}{6}$.

Plug $\left(\frac{1}{2}, -\frac{2\sqrt{3}}{3}\right)$ into $y = -x\sqrt{3} - \frac{\sqrt{3}}{6}$.

Plug $\left(1, -\frac{\sqrt{3}}{6}\right)$ into $y = x\sqrt{3} - \frac{7\sqrt{3}}{6}$.

Plug $\left(1, -\frac{\sqrt{3}}{6}\right)$ into $y = -\frac{x\sqrt{3}}{3} + \frac{\sqrt{3}}{6}$.

Plug $\left(-\frac{1}{2}, \frac{\sqrt{3}}{3}\right)$ into $y = -x\sqrt{3} - \frac{\sqrt{3}}{6}$.

Plug $\left(-\frac{1}{2}, \frac{\sqrt{3}}{3}\right)$ into $y = -\frac{x\sqrt{3}}{3} + \frac{\sqrt{3}}{6}$.

Solution to Exercise #28

(A) $\dfrac{1}{x+\sqrt{x^2+1}} + \left(\dfrac{1}{x+\sqrt{x^2+1}}\right)\left(\dfrac{x}{\sqrt{x^2+1}}\right)$

Make a common denominator.

$$= \frac{1}{x+\sqrt{x^2+1}}\left(\frac{\sqrt{x^2+1}}{\sqrt{x^2+1}}\right) + \left(\frac{1}{x+\sqrt{x^2+1}}\right)\left(\frac{x}{\sqrt{x^2+1}}\right)$$

$$= \frac{\sqrt{x^2+1}+x}{x\sqrt{x^2+1}+x^2+1}$$

To rationalize the denominator, multiply by the conjugate. The conjugate of an expression of the form $t\sqrt{u}+w$ is $-t\sqrt{u}+w$ because the irrational terms will cancel out: $\left(t\sqrt{u}+w\right)\left(-t\sqrt{u}+w\right) = -t^2u + wt\sqrt{u} - wt\sqrt{u} + w^2 = -t^2u + w^2$.

$$= \frac{\sqrt{x^2+1}+x}{x\sqrt{x^2+1}+x^2+1}\left(\frac{-x\sqrt{x^2+1}+x^2+1}{-x\sqrt{x^2+1}+x^2+1}\right)$$

It is simpler to treat (x^2+1) as a single entity.

$$= \frac{-x(x^2+1) + \sqrt{x^2+1}(x^2+1) - x^2\sqrt{x^2+1} + x(x^2+1)}{-x^2(x^2+1) + x\sqrt{x^2+1}(x^2+1) - x\sqrt{x^2+1}(x^2+1) + (x^2+1)^2}$$

$$= \frac{\sqrt{x^2+1}(x^2+1) - x^2\sqrt{x^2+1}}{-x^2(x^2+1) + (x^2+1)^2}$$

Note that $\dfrac{\sqrt{y}}{y} = \dfrac{1}{\sqrt{y}}$ because $\sqrt{y}\sqrt{y} = y$.

$$= \frac{\sqrt{x^2+1}(x^2+1-x^2)}{-x^4-x^2+x^4+2x^2+1} = \frac{\sqrt{x^2+1}}{x^2+1} = \frac{1}{\sqrt{x^2+1}} = \frac{P}{Q^k}$$

Note that $z^{1/2} = \sqrt{z}$ because $z^{1/2}z^{1/2} = z$ and $\sqrt{z}\sqrt{z} = z$.

$$\boxed{P=1} \quad , \quad \boxed{Q=x^2+1} \quad , \quad \boxed{k=\frac{1}{2}}$$

(B) $\dfrac{1}{\sqrt{\left(\sqrt{3}\right)^2+1}} = \dfrac{1}{\sqrt{3+1}} = \dfrac{1}{\sqrt{4}} = \boxed{\dfrac{1}{2}}$

Part B is continued on the next page.

$$\frac{1}{\sqrt{3}+\sqrt{\left(\sqrt{3}\right)^2+1}}+\left[\frac{1}{\sqrt{3}+\sqrt{\left(\sqrt{3}\right)^2+1}}\right]\left[\frac{\sqrt{3}}{\sqrt{\left(\sqrt{3}\right)^2+1}}\right]$$

$$=\frac{1}{\sqrt{3}+\sqrt{3+1}}+\left(\frac{1}{\sqrt{3}+\sqrt{3+1}}\right)\left(\frac{\sqrt{3}}{\sqrt{3+1}}\right)$$

$$=\frac{1}{\sqrt{3}+\sqrt{4}}+\left(\frac{1}{\sqrt{3}+\sqrt{4}}\right)\left(\frac{\sqrt{3}}{\sqrt{4}}\right)$$

$$=\frac{1}{\sqrt{3}+2}+\left(\frac{1}{\sqrt{3}+2}\right)\left(\frac{\sqrt{3}}{2}\right)$$

$$=\frac{1}{\sqrt{3}+2}\left(\frac{2}{2}\right)+\left(\frac{1}{\sqrt{3}+2}\right)\left(\frac{\sqrt{3}}{2}\right)$$

$$=\frac{2+\sqrt{3}}{(\sqrt{3}+2)2}=\boxed{\frac{1}{2}} \quad ✓$$

Solution to Exercise #29	
(A) $2x = 3y$	(B) $2x = wz$

(A) $2x = 3y$

$$\frac{2x}{3} = y$$

$$x + 2 = y + z$$

$$x + 2 = \frac{2x}{3} + z$$

Multiply both sides by 3.

$$3x + 6 = 2x + 3z$$

$$\boxed{x} = 3z - 6 = \boxed{3(z-2)}$$

$$\boxed{y} = \frac{2x}{3} = \frac{2}{3}3(z-2) = \boxed{2(z-2)}$$

(B) $2x = wz$

$$x = \frac{wz}{2}$$

$$x = 3(z-2)$$

$$\frac{wz}{2} = 3(z-2)$$

$$wz = 6(z-2) = 6z - 12$$

$$12 = 6z - wz = z(6-w)$$

$$\boxed{\frac{12}{6-w} = z}$$

(C) $z > 0$

$$z = \frac{12}{6-w}$$

$$6 - w > 0$$

$$6 > w$$

$$w > 0$$

$$\boxed{6 > w > 0}$$

Notes: Exercise 29 is modeled after Euler's formula, $V + F = E + 2$, which relates the number of vertices, faces, and edges of a polyhedron or of a planar graph. The formulas $2E = 3F$, $E = 3(V - 2)$, and $F = 2(V - 2)$ correspond to a polyhedron with triangular faces or to a maximal planar graph. The formula $V = \frac{12}{6-D}$ represents that any maximal planar graph (or polyhedron with triangular faces) has at least one vertex with a degree (D) of 5 or less (where the degree of a vertex equals the number of edges that intersect at the vertex). Since the symbols V and F can effectively be swapped in Euler's formula, a similar set of formulas apply to a dual polyhedron or to a map where every vertex has degree three, such that at least one face must be bounded by 5 or fewer edges.

Solution to Exercise #30

(A) $y = 36x^2 - 216x + 144 + 180 - 180$

$$y = 36x^2 - 216x + 324 - 180$$

$$y = (6x - 18)^2 - 180$$

$$y = (ax + b)^2 + c$$

$\boxed{a = 6}$, $\boxed{b = -18}$, $\boxed{c = -180}$

(B) $y = px^2 + qx + r + \dfrac{q^2}{4p} - \dfrac{q^2}{4p}$

$$y = px^2 + qx + \frac{q^2}{4p} + r - \frac{q^2}{4p}$$

$$y = \left(x\sqrt{p} + \frac{q}{2\sqrt{p}}\right)^2 + r - \frac{q^2}{4p}$$

$$y = (ax + b)^2 + c$$

$\boxed{a = \sqrt{p}}$, $\boxed{b = \dfrac{q}{2\sqrt{p}}}$ or $\boxed{b = \dfrac{q\sqrt{p}}{2p}}$

$\boxed{c = r - \dfrac{q^2}{4p}}$

(C) $y = 36x^2 - 216x + 144$

$p = 36$, $q = -216$, $r = 144$

$$a = \sqrt{p} = \sqrt{36} = \boxed{6} \quad \checkmark$$

$$b = \frac{q}{2\sqrt{p}} = \frac{-216}{2\sqrt{36}} = \frac{-216}{2(6)} = \boxed{-18} \quad \checkmark$$

$$c = r - \frac{q^2}{4p} = 144 - \frac{(-216)^2}{4(36)}$$

$$= 144 - \frac{216}{4}\frac{216}{36}$$

$$c = 144 - 54(6)$$

$$= 144 - 324 = \boxed{-180} \quad \checkmark$$

$$\frac{q}{2\sqrt{p}} = \frac{q}{2\sqrt{p}}\frac{\sqrt{p}}{\sqrt{p}} = \frac{q\sqrt{p}}{2p}$$

Note: This technique of algebra is called "completing the square."

Solution to Exercise #31

(A) $u = \dfrac{x}{p}$

$up = x$

$p = \dfrac{x}{u}$

$w = \dfrac{x}{q}$

$wq = x$

$q = \dfrac{x}{w}$

$y = p + q = \dfrac{x}{u} + \dfrac{x}{w} = x\left(\dfrac{1}{u} + \dfrac{1}{w}\right)$

$\dfrac{y}{x} = \dfrac{1}{u} + \dfrac{1}{w} = \dfrac{w}{uw} + \dfrac{u}{uw} = \dfrac{u+w}{uw}$

$\dfrac{x}{y} = \dfrac{uw}{u+w}$

$\boxed{z} = \dfrac{2x}{y} = \boxed{\dfrac{2uw}{u+w}}$

(B) $z = \dfrac{2uw}{u+w} = \dfrac{2(20)(30)}{20+30} = \dfrac{1200}{50}$

$= \boxed{24}$

$p = \dfrac{x}{u} = \dfrac{240}{20} = \boxed{12}$

$q = \dfrac{x}{w} = \dfrac{240}{30} = \boxed{8}$

$y = p + q = 12 + 8 = \boxed{20}$

$z = \dfrac{2x}{y} = \dfrac{2(240)}{20} = 24$ ✔

Solution to Exercise #32

(A) $\left(\dfrac{1}{2} - \left(2 - \left(1 - \left(3 - \left(\dfrac{1}{x} + \dfrac{1}{3}\right)^{-1}\right)^{-1}\right)^{-1}\right)^{-1}\right)^{-1} = \dfrac{3}{4}$

$\dfrac{1}{2} - \left(2 - \left(1 - \left(3 - \left(\dfrac{1}{x} + \dfrac{1}{3}\right)^{-1}\right)^{-1}\right)^{-1}\right)^{-1} = \dfrac{4}{3}$

$\dfrac{1}{2} = \dfrac{4}{3} + \left(2 - \left(1 - \left(3 - \left(\dfrac{1}{x} + \dfrac{1}{3}\right)^{-1}\right)^{-1}\right)^{-1}\right)^{-1}$

$\dfrac{1}{2} - \dfrac{4}{3} = -\dfrac{5}{6} = \left(2 - \left(1 - \left(3 - \left(\dfrac{1}{x} + \dfrac{1}{3}\right)^{-1}\right)^{-1}\right)^{-1}\right)^{-1}$

Take the reciprocal of both sides of the equation:

$y^{-1} = \dfrac{1}{y} = \dfrac{3}{4}$

$y = \dfrac{4}{3}$

Add the expression in parentheses to both sides. Subtract $\dfrac{4}{3}$.

$\dfrac{1}{2} - \dfrac{4}{3} = \dfrac{3}{6} - \dfrac{8}{6} = -\dfrac{5}{6}$

$$-\frac{6}{5} = 2 - \left(1 - \left(3 - \left(\frac{1}{x} + \frac{1}{3}\right)^{-1}\right)^{-1}\right)^{-1}$$

$$\left(1 - \left(3 - \left(\frac{1}{x} + \frac{1}{3}\right)^{-1}\right)^{-1}\right)^{-1} - \frac{6}{5} = 2$$

$$\left(1 - \left(3 - \left(\frac{1}{x} + \frac{1}{3}\right)^{-1}\right)^{-1}\right)^{-1} = 2 + \frac{6}{5} = \frac{10}{5} + \frac{6}{5} = \frac{16}{5}$$

$$1 - \left(3 - \left(\frac{1}{x} + \frac{1}{3}\right)^{-1}\right)^{-1} = \frac{5}{16}$$

$$1 = \frac{5}{16} + \left(3 - \left(\frac{1}{x} + \frac{1}{3}\right)^{-1}\right)^{-1}$$

$$1 - \frac{5}{16} = \frac{16}{16} - \frac{5}{16} = \frac{11}{16} = \left(3 - \left(\frac{1}{x} + \frac{1}{3}\right)^{-1}\right)^{-1}$$

$$\frac{16}{11} = 3 - \left(\frac{1}{x} + \frac{1}{3}\right)^{-1}$$

$$\frac{16}{11} + \left(\frac{1}{x} + \frac{1}{3}\right)^{-1} = 3$$

$$\left(\frac{1}{x} + \frac{1}{3}\right)^{-1} = 3 - \frac{16}{11} = \frac{33}{11} - \frac{16}{11} = \frac{17}{11}$$

$$\frac{1}{x} + \frac{1}{3} = \frac{11}{17}$$

$$\frac{1}{x} = \frac{11}{17} - \frac{1}{3} = \frac{33}{51} - \frac{17}{51} = \frac{16}{51}$$

$$x = \boxed{\frac{51}{16}} = \boxed{3.1875}$$

$$(B) \quad \left(\frac{1}{2} - \left(2 - \left(1 - \left(3 - \left(\frac{16}{51} + \frac{17}{51}\right)^{-1}\right)^{-1}\right)^{-1}\right)^{-1}\right)^{-1}$$

$$= \frac{3}{4}$$

Take the reciprocal of both sides.

$$-\frac{5}{6} = z^{-1} = \frac{1}{z}$$

$$-\frac{6}{5} = z$$

Continue to isolate the expression in parentheses and take the reciprocal of both sides.

Make a common denominator.

$$\left(\frac{1}{2}-\left(2-\left(1-\left(3-\left(\frac{33}{51}\right)^{-1}\right)^{-1}\right)^{-1}\right)^{-1}\right)^{-1}=\frac{3}{4}$$

Take the reciprocal of $\frac{33}{51}$.

$$\left(\frac{1}{2}-\left(2-\left(1-\left(\frac{99}{33}-\frac{51}{33}\right)^{-1}\right)^{-1}\right)^{-1}\right)^{-1}=\frac{3}{4}$$

Make a common denominator.

$$\left(\frac{1}{2}-\left(2-\left(1-\left(\frac{48}{33}\right)^{-1}\right)^{-1}\right)^{-1}\right)^{-1}=\frac{3}{4}$$

$\frac{48}{33}=\frac{16}{11}$.

$$\left(\frac{1}{2}-\left(2-\left(1-\left(\frac{16}{11}\right)^{-1}\right)^{-1}\right)^{-1}\right)^{-1}=\frac{3}{4}$$

Take the reciprocal of $\frac{16}{11}$.

$$\left(\frac{1}{2}-\left(2-\left(\frac{16}{16}-\frac{11}{16}\right)^{-1}\right)^{-1}\right)^{-1}=\frac{3}{4}$$

Continue to make a common denominator and take reciprocals.

$$\left(\frac{1}{2}-\left(2-\left(\frac{5}{16}\right)^{-1}\right)^{-1}\right)^{-1}=\frac{3}{4}$$

$$\left(\frac{1}{2}-\left(\frac{10}{5}-\frac{16}{5}\right)^{-1}\right)^{-1}=\frac{3}{4}$$

$$\left(\frac{1}{2}-\left(-\frac{6}{5}\right)^{-1}\right)^{-1}=\frac{3}{4}$$

$$\left(\frac{1}{2}-\left(-\frac{5}{6}\right)\right)^{-1}=\frac{3}{4}$$

$$\left(\frac{3}{6}+\frac{5}{6}\right)^{-1}=\left(\frac{8}{6}\right)^{-1}=\frac{6}{8}=\frac{3}{4}\quad\checkmark$$

Solution to Exercise #33

$$(A) \ (x + a)(x^2 + bx + cx + bc) = x^3 + px^2 + qx + r$$

$$x^3 + bx^2 + cx^2 + bcx + ax^2 + abx + acx + abc = x^3 + px^2 + qx + r$$

Factor out x^2. Also factor out x.

$$x^3 + (a + b + c)x^2 + (ab + ac + bc)x + abc = x^3 + px^2 + qx + r$$

Subtract $x^3 + px^2 + qx + r$ from both sides of the equation; x^3 cancels out.

$$(a + b + c - p)x^2 + (ab + ac + bc - q)x + (abc - r) = 0$$

$$a + b + c - p = 0$$

$$\boxed{a + b + c = p}$$

$$abc - r = 0$$

$$\boxed{abc = r}$$

$$ab + ac + bc - q = 0$$

$$ab + ac + bc = q$$

$$\frac{ab}{abc} + \frac{ac}{abc} + \frac{bc}{abc} = \frac{q}{abc}$$

$$\boxed{\frac{1}{c} + \frac{1}{b} + \frac{1}{a} = \frac{q}{r}}$$

$$(B) \ a = 2, b = 3, c = 4, x = 5$$

$$p = a + b + c = 2 + 3 + 4 = 9$$

$$r = abc = (2)(3)(4) = 24$$

$$\frac{q}{r} = \frac{1}{c} + \frac{1}{b} + \frac{1}{a} = \frac{1}{4} + \frac{1}{3} + \frac{1}{2}$$

$$\frac{q}{r} = \frac{3}{12} + \frac{4}{12} + \frac{6}{12} = \frac{13}{12}$$

$$q = \frac{13}{12}r = \frac{13}{12}(24) = 26$$

$$(x + a)(x + b)(x + c) = (5 + 2)(5 + 3)(5 + 4)$$

$$= (7)(8)(9) = (7)(72) = \boxed{504}$$

$$x^3 + px^2 + qx + r = 5^3 + 9(5)^2 + 26(5) + 24$$

$$= 125 + 9(25) + 130 + 24 = 279 + 225 = \boxed{504} \quad \checkmark$$

Solution to Exercise #34

(A) $\left(\sqrt{x} - 1 + \dfrac{1}{\sqrt{x}}\right)^2 = \sqrt{x}\sqrt{x} - \sqrt{x}(1) + \dfrac{\sqrt{x}}{\sqrt{x}} - 1\sqrt{x} + 1 - \dfrac{1}{\sqrt{x}} + \dfrac{\sqrt{x}}{\sqrt{x}} - \dfrac{1}{\sqrt{x}} + \dfrac{1}{\sqrt{x}\sqrt{x}}$

$$= x - \sqrt{x} + 1 - \sqrt{x} + 1 - \dfrac{1}{\sqrt{x}} + 1 - \dfrac{1}{\sqrt{x}} + \dfrac{1}{x}$$

$$= x - 2\sqrt{x} + 3 - \dfrac{2}{\sqrt{x}} + \dfrac{1}{x}$$

$$\left(\sqrt{x} - 1 + \dfrac{1}{\sqrt{x}}\right)^3 = \left(\sqrt{x} - 1 + \dfrac{1}{\sqrt{x}}\right)\left(\sqrt{x} - 1 + \dfrac{1}{\sqrt{x}}\right)^2$$

$$= \left(\sqrt{x} - 1 + \dfrac{1}{\sqrt{x}}\right)\left(x - 2\sqrt{x} + 3 - \dfrac{2}{\sqrt{x}} + \dfrac{1}{x}\right)$$

$$= x\sqrt{x} - 2\sqrt{x}\sqrt{x} + 3\sqrt{x} - \dfrac{2\sqrt{x}}{\sqrt{x}} + \dfrac{\sqrt{x}}{x} - x + 2\sqrt{x} - 3 + \dfrac{2}{\sqrt{x}} - \dfrac{1}{x}$$

$$+ \dfrac{x}{\sqrt{x}} - \dfrac{2\sqrt{x}}{\sqrt{x}} + \dfrac{3}{\sqrt{x}} - \dfrac{2}{\sqrt{x}\sqrt{x}} + \dfrac{1}{x\sqrt{x}}$$

$$= x\sqrt{x} - 2x + 3\sqrt{x} - 2 + \dfrac{1}{\sqrt{x}} - x + 2\sqrt{x} - 3 + \dfrac{2}{\sqrt{x}} - \dfrac{1}{x} + \sqrt{x} - 2 + \dfrac{3}{\sqrt{x}} - \dfrac{2}{x} + \dfrac{1}{x\sqrt{x}}$$

$$= x\sqrt{x} - 3x + 6\sqrt{x} - 7 + \dfrac{6}{\sqrt{x}} - \dfrac{3}{x} + \dfrac{1}{x\sqrt{x}}$$

$$= x\sqrt{x} - 3x + 6\sqrt{x} - 7 + \dfrac{6}{\sqrt{x}}\dfrac{\sqrt{x}}{\sqrt{x}} - \dfrac{3}{x} + \dfrac{1}{x\sqrt{x}}\dfrac{\sqrt{x}}{\sqrt{x}}$$

$$= x\sqrt{x} - 3x + 6\sqrt{x} - 7 + \dfrac{6\sqrt{x}}{x} - \dfrac{3}{x} + \dfrac{\sqrt{x}}{x^2}$$

$$= \left(x + 6 + \dfrac{6}{x} + \dfrac{1}{x^2}\right)\sqrt{x} - 3x - 7 - \dfrac{3}{x}$$

$$= (x^3 + 6x^2 + 6x + 1)\dfrac{\sqrt{x}}{x^2} + (-3x^3 - 7x^2 - 3x)\dfrac{1}{x^2}$$

$$= (P\sqrt{x} + Q)\dfrac{1}{x^2} = P\dfrac{\sqrt{x}}{x^2} + Q\dfrac{1}{x^2}$$

$$P = \boxed{x^3 + 6x^2 + 6x + 1}$$

$$Q = \boxed{-3x^3 - 7x^2 - 3x}$$

(B) $P = x^3 + 6x^2 + 6x + 1 = 4^3 + 6(4)^2 + 6(4) + 1$

$$P = 64 + 6(16) + 24 + 1 = 89 + 96 = 185$$

$$Q = -3x^3 - 7x^2 - 3x = -3(4)^3 - 7(4)^2 - 3(4)$$

$$Q = -3(64) - 7(16) - 12 = -192 - 112 - 12 = -316$$

$$(P\sqrt{x} + Q)\frac{1}{x^2} = \frac{185\sqrt{4} - 316}{4^2} = \frac{185(2) - 316}{16} = \frac{370 - 316}{16} = \frac{54}{16} = \boxed{\frac{27}{8}}$$

$$\left(\sqrt{x} - 1 + \frac{1}{\sqrt{x}}\right)^3 = \left(\sqrt{4} - 1 + \frac{1}{\sqrt{4}}\right)^3 = \left(2 - 1 + \frac{1}{2}\right)^3$$

$$= \left(1 + \frac{1}{2}\right)^3 = \left(\frac{3}{2}\right)^3 = \frac{3^3}{2^3} = \boxed{\frac{27}{8}} \quad \checkmark$$

(C) $x = 0$ would cause division by zero.

Solution to Exercise #35

(A) $xz = 9$, $xy = 16$

Divide the two equations above.

$$\frac{z}{y} = \frac{9}{16}$$

$$yz = 25$$

Multiply the two equations above.

$$z^2 = \frac{9(25)}{16}$$

$$z = \pm\sqrt{\frac{9(25)}{16}} = \pm\frac{\sqrt{9}\sqrt{25}}{\sqrt{16}}$$

$$z = \pm\frac{(3)(5)}{4} = \boxed{\pm\frac{15}{4}}$$

$$yz = 25 \quad \to \quad y = \frac{25}{z} = \frac{25}{\pm 15/4}$$

$$y = \pm\frac{(4)(25)}{15} = \pm\frac{100}{15} = \boxed{\pm\frac{20}{3}}$$

$$x = \frac{9}{z} = \frac{9}{\pm 15/4} = \pm\frac{(9)(4)}{15} = \boxed{\pm\frac{12}{5}}$$

(B) $\dfrac{9}{z} = \dfrac{9}{\left(\pm\dfrac{15}{4}\right)} = 9\left(\pm\dfrac{4}{15}\right)$

$$= \pm\frac{36}{15} = \pm\frac{12}{5} = x \quad \checkmark$$

$$\frac{16}{x} = \frac{16}{\left(\pm\dfrac{12}{5}\right)} = 16\left(\pm\frac{5}{12}\right)$$

$$= \pm\frac{80}{12} = \pm\frac{20}{3} = y \quad \checkmark$$

$$\frac{25}{y} = \frac{25}{\left(\pm\dfrac{20}{3}\right)} = 25\left(\pm\frac{3}{20}\right)$$

$$= \pm\frac{75}{20} = \pm\frac{15}{4} = z \quad \checkmark$$

Either x, y, and z are all positive or they are all negative.

Solution to Exercise #36

(A)
$$\frac{y\left(\frac{q}{p}\right)^{\frac{1-k}{k}} - x\left[(p)^{\frac{1-k}{k}}\right]}{y - x\left(\frac{p}{q}\right)^{\frac{1-k}{k}}\left[(p)^{\frac{1-k}{k}}\right]}$$

$$= \frac{y(q)^{\frac{1-k}{k}} - x(p)^{\frac{1-k}{k}}\left[\frac{1}{(p)^{\frac{1-k}{k}}}\right]}{y - x\left(\frac{p}{q}\right)^{\frac{1-k}{k}}}$$

$$= \frac{y(q)^{\frac{1-k}{k}} - x(p)^{\frac{1-k}{k}}\left[\frac{1}{(p)^{\frac{1-k}{k}}}\right]\left[(q)^{\frac{1-k}{k}}\right]}{y - x\left(\frac{p}{q}\right)^{\frac{1-k}{k}}\left[\frac{1}{(p)^{\frac{1-k}{k}}}\right]\left[(q)^{\frac{1-k}{k}}\right]}$$

$$= \frac{y(q)^{\frac{1-k}{k}} - x(p)^{\frac{1-k}{k}}\left[(q)^{\frac{1-k}{k}}\right]}{y(q)^{\frac{1-k}{k}} - x(p)^{\frac{1-k}{k}}\left[(p)^{\frac{1-k}{k}}\right]}$$

$$= \left[\frac{(q)^{\frac{1-k}{k}}}{(p)^{\frac{1-k}{k}}}\right] = \left(\frac{q}{p}\right)^{\frac{1-k}{k}} = r^c$$

$$\boxed{r = \frac{q}{p}} \quad , \quad \boxed{c = \frac{1-k}{k}}$$

(B) $k = 0$, $p = 0$, $q = 0$

and $y = x\left(\frac{p}{q}\right)^{\frac{1-k}{k}}$

(C) $r = \frac{q}{p} = \frac{32}{4} = 8$

$$c = \frac{1 - \frac{3}{2}}{\frac{3}{2}} = \frac{-\frac{1}{2}}{\frac{3}{2}} = -\frac{1}{2} \div \frac{3}{2}$$

$$= -\frac{1}{2} \times \frac{2}{3} = -\frac{1}{3}$$

$$r^c = 8^{-1/3} = \frac{1}{8^{1/3}} = \frac{1}{\sqrt[3]{8}} = \boxed{\frac{1}{2}}$$

Multiply by $\dfrac{(p)^{\frac{1-k}{k}}}{(p)^{\frac{1-k}{k}}}$ (which equals one).

Distribute $(p)^{\frac{1-k}{k}}$ in the numerator (but not the denominator):

$$(a - b)c = ac - bc$$

Since $\dfrac{(p)^{\frac{1-k}{k}}}{(p)^{\frac{1-k}{k}}} = 1$, there is cancellation:

$$\left(\frac{q}{p}\right)^{\frac{1-k}{k}}(p)^{\frac{1-k}{k}} = (q)^{\frac{1-k}{k}}$$

Multiply by $\dfrac{(q)^{\frac{1-k}{k}}}{(q)^{\frac{1-k}{k}}}$. This time, distribute

in the denominator (not the numerator).

$$\left(\frac{p}{q}\right)^{\frac{1-k}{k}}(q)^{\frac{1-k}{k}} = (p)^{\frac{1-k}{k}}$$

There is cancellation on the left:

$$\frac{y(q)^{\frac{1-k}{k}} - x(p)^{\frac{1-k}{k}}}{y(q)^{\frac{1-k}{k}} - x(p)^{\frac{1-k}{k}}} = 1$$

Apply the rule $\left(\dfrac{u}{w}\right)^m = \dfrac{u^m}{w^m}$:

$$\left[\frac{(q)^{\frac{1-k}{k}}}{(p)^{\frac{1-k}{k}}}\right] = \left(\frac{q}{p}\right)^{\frac{1-k}{k}}$$

(B) The four values listed would cause division by zero.

Note: Similar algebra arises in thermodynamics in the context of heat engines. Although the algebra in this problem bears some measure of resemblance to the efficiency of the Brayton cycle, r in this problem is the reciprocal of the pressure ratio that appears in the Brayton cycle; also, the expression given in the problem would be subtracted from one to find the efficiency.

Solution to Exercise #37

(A) $w(ckx - z) = ky^2$

$2ckw(1 - x) = ky^2$

$w(ckx - z) = 2ckw(1 - x)$

$ckx - z = 2ck(1 - x)$

$ckx - z = 2ck - 2ckx$

$3ckx - 2ck = z$

$\boxed{ck(3x - 2) = z}$

(B) $3x - 2 = 0$

$3x = 2$

$\boxed{x = \dfrac{2}{3}}$

Note: The math for Parts A and B is similar to a classic conservation of energy problem in physics with an ice cube sliding along a frictionless hemispherical dome. In physics, the equations look like $mg \cos \theta - N = \dfrac{mv^2}{R}$ and $mgR(1 - \cos \theta) = \dfrac{mv^2}{2}$. In this problem, x plays the role of $\cos \theta$. When $x = \cos \theta = \dfrac{2}{3}$, the ice cube loses contact with the hemisphere. Of course, it is NOT necessary to know any physics (or trig) to solve this algebra problem.

Solution to Exercise #38

(A) $m = \dfrac{y_2 - y_1}{x_2 - x_1} = \dfrac{\dfrac{1}{2} - \left(-\dfrac{1}{2}\right)}{\dfrac{3\sqrt{3}}{4} - \left(-\dfrac{\sqrt{3}}{4}\right)}$

$m = \dfrac{\dfrac{1}{2} + \dfrac{1}{2}}{\dfrac{3\sqrt{3}}{4} + \dfrac{\sqrt{3}}{4}} = \dfrac{1}{\sqrt{3}} = \dfrac{1}{\sqrt{3}} \dfrac{\sqrt{3}}{\sqrt{3}} = \dfrac{\sqrt{3}}{3}$

$y = mx + b = \dfrac{\sqrt{3}}{3} x + b$

Find the slope from $\left(-\dfrac{\sqrt{3}}{4}, -\dfrac{1}{2}\right)$ to $\left(\dfrac{3\sqrt{3}}{4}, \dfrac{1}{2}\right)$. Subtracting a negative number equates to addition.

$\dfrac{3\sqrt{3}}{4} + \dfrac{\sqrt{3}}{4} = \dfrac{4\sqrt{3}}{4} = \sqrt{3}$

Rationalize the denominator:

$\dfrac{1}{\sqrt{3}} = \dfrac{1}{\sqrt{3}} \dfrac{\sqrt{3}}{\sqrt{3}} = \dfrac{\sqrt{3}}{3}$

$$\frac{1}{2} = \frac{\sqrt{3}}{3}\left(\frac{3\sqrt{3}}{4}\right) + b$$

$$\frac{1}{2} = \frac{\sqrt{3}\sqrt{3}}{4} + b$$

$$\frac{1}{2} = \frac{3}{4} + b$$

$$\frac{1}{2} - \frac{3}{4} = \frac{2}{4} - \frac{3}{4} = -\frac{1}{4} = b$$

$$\boxed{y = \frac{\sqrt{3}}{3}x - \frac{1}{4}}$$

Plug $m = \frac{\sqrt{3}}{3}$ and $\left(\frac{3\sqrt{3}}{4}, \frac{1}{2}\right)$ into $y = mx + b$. The 3 cancels out. Recall that $\sqrt{3}\sqrt{3} = 3$.

(B) $m_\perp = -\frac{1}{m} = -\frac{1}{\sqrt{3}/3} = -\frac{3}{\sqrt{3}} = -\sqrt{3}$

$$x_{ave} = \frac{x_1 + x_2}{2} = \frac{\frac{3\sqrt{3}}{4} - \frac{\sqrt{3}}{4}}{2} = \frac{\frac{\sqrt{3}}{2}}{2}$$

$$= \frac{\sqrt{3}}{2} \div 2 = \frac{\sqrt{3}}{2} \times \frac{1}{2} = \frac{\sqrt{3}}{4}$$

$$y_{ave} = \frac{y_1 + y_2}{2} = \frac{\frac{1}{2} - \frac{1}{2}}{2} = \frac{0}{2} = 0$$

$$y_{ave} = m_\perp x_{ave} + b_\perp$$

$$0 = -\sqrt{3}\left(\frac{\sqrt{3}}{4}\right) + b_\perp$$

$$0 = -\frac{3}{4} + b_\perp$$

$$\frac{3}{4} = b_\perp$$

$$\boxed{y_\perp = -x\sqrt{3} + \frac{3}{4}}$$

The negative reciprocal of the slope equals the slope of the perpendicular line.

The coordinates of the midpoint can be found by averaging the x-coordinates and y-coordinates of the given points. The coordinates of the midpoint are $\left(\frac{\sqrt{3}}{4}, 0\right)$.

Plug $m_\perp = -\sqrt{3}$ and $\left(\frac{\sqrt{3}}{4}, 0\right)$ into $y = mx + b$.

(C) $\frac{1}{2} = \frac{\sqrt{3}}{3}\left(\frac{3\sqrt{3}}{4}\right) - \frac{1}{4} = \frac{\sqrt{3}\sqrt{3}}{4} - \frac{1}{4}$

$$= \frac{3}{4} - \frac{1}{4} = \frac{2}{4} = \frac{1}{2} \quad \checkmark$$

Plug $\left(\frac{3\sqrt{3}}{4}, \frac{1}{2}\right)$ into $y = \frac{\sqrt{3}}{3}x - \frac{1}{4}$.

$-\dfrac{1}{2} = \dfrac{\sqrt{3}}{3}\left(-\dfrac{\sqrt{3}}{4}\right) - \dfrac{1}{4} = -\dfrac{3}{3(4)} - \dfrac{1}{4}$	Plug $\left(-\dfrac{\sqrt{3}}{4}, -\dfrac{1}{2}\right)$ into $y = \dfrac{\sqrt{3}}{3}x - \dfrac{1}{4}$.
$= -\dfrac{1}{4} - \dfrac{1}{4} = -\dfrac{2}{4} = -\dfrac{1}{2}$ ✔	Both given points and the midpoint pass through $y = \dfrac{\sqrt{3}}{3}x - \dfrac{1}{4}$.
$0 = \dfrac{\sqrt{3}}{3}\left(\dfrac{\sqrt{3}}{4}\right) - \dfrac{1}{4} = \dfrac{3}{3(4)} - \dfrac{1}{4}$	Plug $\left(\dfrac{\sqrt{3}}{4}, 0\right)$ into $y = \dfrac{\sqrt{3}}{3}x - \dfrac{1}{4}$.
$= \dfrac{1}{4} - \dfrac{1}{4} = 0$ ✔	Only the midpoint passes through the perpendicular line.
$0 = -\dfrac{\sqrt{3}}{4}\sqrt{3} + \dfrac{3}{4} = -\dfrac{3}{4} + \dfrac{3}{4} = 0$ ✔	Plug $\left(\dfrac{\sqrt{3}}{4}, 0\right)$ into $y = -x\sqrt{3} + \dfrac{3}{4}$.

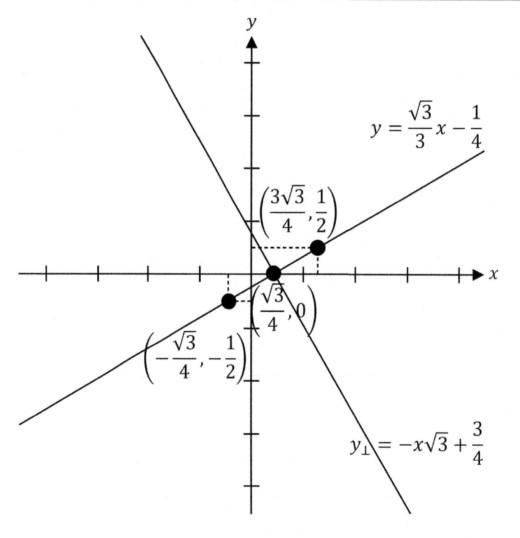

Solution to Exercise #39	
(A) $45 = 50x - 5x^2$ $5x^2 - 50x + 45 = 0$ $x^2 - 10 + 9 = 0$ $(x-1)(x-9) = 0$ $x = 1 \quad$ or $\quad x = 9$ $z = 50x\sqrt{3} = 50(1)\sqrt{3} \quad$ or $\quad 50(9)\sqrt{3}$ $z = \boxed{50\sqrt{3}} \quad$ or $\quad z = \boxed{450\sqrt{3}}$ (B) $50\sqrt{3} = 50x\sqrt{3}$ $\dfrac{50\sqrt{3}}{50\sqrt{3}} = 1 = x$ $y = 50x - 5x^2 = 50(1) - 5(1)^2$ $y = 50 - 5(1) = 50 - 5 = 45 \quad \checkmark$ $450\sqrt{3} = 50x\sqrt{3}$ $\dfrac{450\sqrt{3}}{50\sqrt{3}} = 9 = x$ $y = 50x - 5x^2 = 50(9) - 5(9)^2$ $y = 450 - 5(81) = 450 - 405 = 45 \quad \checkmark$	(C) $\dfrac{y}{x} = 15$ $y = 15x$ $y = 50x - 5x^2$ $15x = 50x - 5x^2$ $5x^2 - 50x + 15x = 0$ $5x^2 - 35x = 0$ $5x(x-7) = 0$ $5x = 0 \quad$ or $\quad x - 7 = 0$ $x = 0 \quad$ or $\quad x = 7$ $x > 0 \Rightarrow x = \boxed{7}$ only $y = 15x = 15(7) = \boxed{105}$ $z = 50x\sqrt{3} = 50(7)\sqrt{3} = \boxed{350\sqrt{3}}$ (D) $y = 50x - 5x^2 = 50(7) - 5(7)^2$ $y = 350 - 5(49) = 350 - 245 = 105 \quad \checkmark$ $z = 50x\sqrt{3} = 50(7)\sqrt{3} = 350\sqrt{3} \quad \checkmark$ $\dfrac{y}{x} = \dfrac{105}{7} = 15 \quad \checkmark$

Solution to Exercise #40	
$q = \dfrac{2}{5}pr^2$ $pcx = \dfrac{1}{2}py^2 + \dfrac{1}{2}qz^2$ $pcx = \dfrac{1}{2}py^2 + \dfrac{1}{2}\left(\dfrac{2}{5}pr^2\right)z^2$ $pcx = \dfrac{1}{2}py^2 + \dfrac{1}{5}pr^2z^2$ $pcx = \dfrac{1}{2}py^2 + \dfrac{1}{5}p(rz)^2$ $y = rz$	Replace q with $\dfrac{2}{5}pr^2$ in the equation $pcx = \dfrac{1}{2}py^2 + \dfrac{1}{2}qz^2$ Note that $\dfrac{1}{2}\dfrac{2}{5} = \dfrac{2}{10} = \dfrac{1}{5}$. Note that $r^2z^2 = (rz)^2$. Since $y = rz$, the expression $(rz)^2$ may be replaced by y^2.

<table>
<tr><td>

$$pcx = \frac{1}{2}py^2 + \frac{1}{5}py^2$$

$$pcx = \frac{5}{10}py^2 + \frac{2}{10}py^2 = \frac{7}{10}py^2$$

$$cx = \frac{7y^2}{10}$$

$$\frac{10cx}{7} = y^2$$

$$\boxed{\sqrt{\frac{10cx}{7}} = y} \text{ or } \boxed{y = \frac{\sqrt{70cx}}{7}}$$

</td><td>

Combine like terms. Make a common denominator.

Divide both sides by p. (Since $p > 0$, there is no danger of dividing by zero. If $p = 0$ were a possibility, the proper way to proceed would be to factor p out.)

Since $y > 0$, only the negative root applies. Note: $\sqrt{\frac{10cx}{7}} = \frac{\sqrt{10cx}}{\sqrt{7}}\frac{\sqrt{7}}{\sqrt{7}} = \frac{\sqrt{70cx}}{7}$.

</td></tr>
<tr><td colspan="2">

Note: This problem is similar to a classic rolling without slipping problem in physics. In physics, the equations for a solid sphere rolling without slipping down an incline from rest look like $mgh = \frac{1}{2}mv^2 + \frac{1}{2}I\omega^2$, $I = \frac{2}{5}mr^2$, and $v = r\omega$.

</td></tr>
</table>

<table>
<tr><td colspan="2" align="center">Solution to Exercise #41</td></tr>
<tr><td>

$$\text{(A) } \frac{24}{x^2} = \frac{96}{y^2}$$

$$24y^2 = 96x^2$$

$$y^2 = 4x^2$$

$$\sqrt{y^2} = \pm\sqrt{4x^2}$$

$$y = \pm2x$$

$$x + y = 18$$

$$x \pm 2x = 18$$

$$x + 2x = 18 \quad \text{or} \quad x - 2x = 18$$

$$3x = 18 \quad \text{or} \quad -x = 18$$

$$x = \frac{18}{3} = \boxed{6} \quad \text{or} \quad x = \boxed{-18}$$

$$x + y = 18$$

$$y = 18 - x$$

$$y = 18 - 6 \quad \text{or} \quad y = 18 - (-18)$$

$$y = \boxed{12} \quad \text{or} \quad y = \boxed{36}$$

</td><td>

Cross multiply:

$$\frac{a}{b} = \frac{c}{d}$$

$$ad = bc$$

Divide both sides by 24.

Square root both sides. Allow for both positive and negative roots. Note that $(-2)^2 = 4$ in addition to $2^2 = 4$.

This leads to two solutions for x and two solutions for y.

One solution is $x = 6$ and $y = 12$. The other solution is $x = -18$ and $y = 36$.

</td></tr>
</table>

$$(B) \quad \frac{24}{6^2} = \frac{96}{12^2}$$

$$\frac{24}{36} = \frac{96}{144}$$

$$\frac{2}{3} = \frac{2}{3} \quad \checkmark$$

$$\frac{24}{(-18)^2} = \frac{96}{36^2}$$

$$\frac{24}{324} = \frac{96}{1296}$$

$$\frac{2}{27} = \frac{2}{27} \quad \checkmark$$

$$6 + 12 = 18 \quad \checkmark$$

$$-18 + 36 = 18 \quad \checkmark$$

To see that $\frac{24}{36}$ reduces to $\frac{2}{3}$, divide the numerator and denominator each by 12.

To see that $\frac{96}{144}$ reduces to $\frac{2}{3}$, divide the numerator and denominator each by 48.

To see that $\frac{24}{324}$ reduces to $\frac{2}{27}$, divide the numerator and denominator each by 12.

To see that $\frac{96}{1296}$ reduces to $\frac{2}{27}$, divide the numerator and denominator each by 48.

Solution to Exercise #42

(A) $\dfrac{72x^5y^2 + 12x^4y^2 - 144x^3y^2}{15x^2y - 20xy}$

$= \dfrac{12x^3y^2(6x^2 + x - 12)}{5xy(3x - 4)}$

$= \dfrac{12x^2y(2x + 3)(3x - 4)}{5(3x - 4)}$

$= \boxed{\dfrac{12x^2y(2x + 3)}{5}} = \boxed{\dfrac{24x^3y + 36x^2y}{5}}$

(B) $\dfrac{12(2)^2(3)[2(2) + 3]}{5}$

$= \dfrac{36(4)(4 + 3)}{5} = \dfrac{144(7)}{5} = \boxed{\dfrac{1008}{5}}$

$\dfrac{72(2)^5(3)^2 + 12(2)^4(3)^2 - 144(2)^3(3)^2}{15(2)^2(3) - 20(2)(3)}$

$= \dfrac{72(32)(9) + 12(16)(9) - 144(8)(9)}{45(4) - 120}$

$= \dfrac{20{,}736 + 1728 - 10{,}368}{180 - 120}$

$= \dfrac{12{,}096}{60} = \boxed{\dfrac{1008}{5}} \quad \checkmark$

(C) Either $x = 0$ or $y = 0$ would cause division by zero. If $x = \frac{4}{3}$, this would make $3x - 4 = 0$, which would also cause division by zero; it would create the indeterminate form of zero divided by zero, since $6x^2 + x - 12$ would also be zero, since $6x^2 + x - 12 = (2x + 3)(3x - 4)$.

Solution to Exercise #43

(A) $\left(x - \dfrac{1}{2}\right)^2 + x^2 + \left(x + \dfrac{3}{2}\right)^2 = \dfrac{15}{2}$

$$x^2 - x + \dfrac{1}{4} + x^2 + x^2 + 3x + \dfrac{9}{4} = \dfrac{15}{2}$$

$$3x^2 + 2x + \dfrac{10}{4} = \dfrac{15}{2}$$

$$3x^2 + 2x + \dfrac{5}{2} = \dfrac{15}{2}$$

$$3x^2 + 2x = \dfrac{15}{2} - \dfrac{5}{2} = \dfrac{10}{2} = 5$$

$$3x^2 + 2x - 5 = 0$$

$$x = \dfrac{-2 \pm \sqrt{2^2 - 4(3)(-5)}}{2(3)}$$

$$x = \dfrac{-2 \pm \sqrt{4 + 60}}{6}$$

$$x = \dfrac{-2 \pm \sqrt{64}}{6} = \dfrac{-2 \pm 8}{6}$$

$$x = \dfrac{-2 - 8}{6} \quad \text{or} \quad x = \dfrac{-2 + 8}{6}$$

$$x = \dfrac{-10}{6} \quad \text{or} \quad x = \dfrac{6}{6}$$

$$x = \boxed{-\dfrac{5}{3}} \quad \text{or} \quad \boxed{x = 1}$$

(B) $\left(-\dfrac{5}{3} - \dfrac{1}{2}\right)^2 + \left(-\dfrac{5}{3}\right)^2 + \left(-\dfrac{5}{3} + \dfrac{3}{2}\right)^2$

$$= \left(-\dfrac{10}{6} - \dfrac{3}{6}\right)^2 + \dfrac{25}{9} + \left(-\dfrac{10}{6} + \dfrac{9}{6}\right)^2$$

$$= \left(-\dfrac{13}{6}\right)^2 + \dfrac{25}{9} + \left(-\dfrac{1}{6}\right)^2$$

$$= \dfrac{169}{36} + \dfrac{25}{9} + \dfrac{1}{36}$$

$$= \dfrac{170}{36} + \dfrac{100}{36} = \dfrac{270}{36} = \dfrac{15}{2} \quad \checkmark$$

$$\left(1 - \dfrac{1}{2}\right)^2 + 1^2 + \left(1 + \dfrac{3}{2}\right)^2$$

$$= \left(\dfrac{2}{2} - \dfrac{1}{2}\right)^2 + 1 + \left(\dfrac{2}{2} + \dfrac{3}{2}\right)^2$$

$$= \left(\dfrac{1}{2}\right)^2 + \dfrac{4}{4} + \left(\dfrac{5}{2}\right)^2$$

$$= \dfrac{1}{4} + \dfrac{4}{4} + \dfrac{25}{4} = \dfrac{30}{4} = \dfrac{15}{2} \quad \checkmark$$

Solution to Exercise #44

(A) $zw^2x^3 = 2^{30}$

$$z = \dfrac{2^{30}}{w^2x^3}$$

$$yz^2w^3 = 2^{26}$$

$$y\left(\dfrac{2^{30}}{w^2x^3}\right)^2 w^3 = 2^{26}$$

Divide both sides by w^2x^3.

Replace z with $\dfrac{2^{30}}{w^2x^3}$ in $yz^2w^3 = 2^{26}$.

Apply the rule $(t^a)^b = t^{ab}$.

$(2^{30})^2 = 2^{60}$, $(w^2)^2 = w^4$, and $(x^3)^2 = x^6$.

$$\frac{2^{60}yw^3}{w^4x^6} = 2^{26}$$

$$\frac{y}{wx^6} = 2^{-34}$$

$$y = 2^{-34}wx^6$$

$$xy^2z^3 = 2^{14}$$

$$x(2^{-34}wx^6)^2\left(\frac{2^{30}}{w^2x^3}\right)^3 = 2^{14}$$

$$x2^{-68}w^2x^{12}\frac{2^{90}}{w^6x^9} = 2^{14}$$

$$\frac{2^{22}x^4}{w^4} = 2^{14}$$

$$x^4 = 2^{-8}w^4$$

$$x = \pm(2^{-8}w^4)^{1/4} = \pm2^{-2}w$$

$$wx^2y^3z^4 = 2^{30}$$

$$w(\pm2^{-2}w)^2(2^{-34}wx^6)^3\left(\frac{2^{30}}{w^2x^3}\right)^4 = 2^{30}$$

$$w2^{-4}w^22^{-102}w^3x^{18}\frac{2^{120}}{w^8x^{12}} = 2^{30}$$

$$\frac{2^{14}x^6}{w^2} = 2^{30}$$

$$\frac{2^{14}(\pm2^{-2}w)^6}{w^2} = 2^{30}$$

$$\frac{2^{14}2^{-12}w^6}{w^2} = 2^{30}$$

$$2^2w^4 = 2^{30}$$

$$w^4 = 2^{28}$$

$$w = \pm(2^{28})^{1/4} = \boxed{\pm2^7} = \boxed{\pm128}$$

$$x = \pm2^{-2}w = \pm2^{-2}(\pm2^7) = \boxed{\pm2^5} = \boxed{\pm32}$$

$$y = 2^{-34}wx^6 = 2^{-34}(\pm2^7)(\pm2^5)^6$$

$$= \pm2^{-34}2^72^{30} = \pm2^{-34}2^{37} = \boxed{\pm2^3} = \boxed{\pm8}$$

$$\frac{w^3}{w^4} = w^{3-4} = w^{-1} = \frac{1}{w}$$

$$\frac{2^{26}}{2^{60}} = 2^{26-60} = 2^{-34}$$

Replace y with $2^{-34}wx^6$ and replace z with $\frac{2^{30}}{w^2x^3}$ in $xy^2z^3 = 2^{14}$.

Apply the rule $(t^a)^b = t^{ab}$. For example, $(2^{-34})^2 = 2^{-68}$ and $(2^{30})^3 = 2^{90}$.

Apply the rules $t^mt^n = t^{m+n}$ and $\frac{t^m}{t^n} = t^{m-n}$.

Raise both sides to the power of $\frac{1}{4}$ because $(x^4)^{1/4} = x^1 = x$. Apply the rule $(t^a)^b = t^{ab}$ to get $(2^{-8})^{1/4} = 2^{-2}$ and $(w^4)^{1/4} = w^1 = w$. Note that the fourth root can be positive or negative.

Plug $x = \pm2^{-2}w$, $y = 2^{-34}wx^6$, and $z = \frac{2^{30}}{w^2x^3}$ into $wx^2y^3z^4 = 2^{30}$.

Apply the rule $(t^a)^b = t^{ab}$. Apply the rules $t^mt^n = t^{m+n}$ and $\frac{t^m}{t^n} = t^{m-n}$.

Raise both sides to the power of $\frac{1}{4}$ because $(w^4)^{1/4} = w^1 = w$. Note that the fourth root can be positive or negative.

Plug $w = \pm2^7$ into $x = \pm2^{-2}w$.

Plug $w = \pm2^7$ and $x = \pm2^5$ into $y = 2^{-34}wx^6$.

$$z = \frac{2^{30}}{w^2 x^3} = \frac{2^{30}}{(\pm 2^7)^2 (\pm 2^5)^3}$$

$$= \pm \frac{2^{30}}{2^{14} 2^{15}} = \pm \frac{2^{30}}{2^{29}} = \boxed{\pm 2^1} = \boxed{\pm 2}$$

$w = 128, x = 32, y = 8,$ and $z = 2$;

or $w = -128, x = 32, y = -8,$ and $z = 2$;

or $w = 128, x = -32, y = 8,$ and $z = -2$;

or $w = -128, x = -32, y = -8,$ and

$$z = -2$$

(B) $wx^2 y^3 z^4 = \pm 2^7 (2^5)^2 (\pm 2^3)^3 (2^1)^4$

$$= 2^7 2^{10} 2^9 2^4 = 2^{30} \quad \checkmark$$

$xy^2 z^3 = \pm 2^5 (2^3)^2 (\pm 2^1)^3$

$$= 2^5 2^6 2^3 = 2^{14} \quad \checkmark$$

$yz^2 w^3 = \pm 2^3 (2^1)^2 (\pm 2^7)^3$

$$= 2^3 2^2 2^{21} = 2^{26} \quad \checkmark$$

$zw^2 x^3 = \pm 2^1 (2^7)^2 (\pm 2^5)^3$

$$= 2^1 2^{14} 2^{15} = 2^{30} \quad \checkmark$$

Plug $w = \pm 2^7$ and $x = \pm 2^5$ into

$$z = \frac{2^{30}}{w^2 x^3}.$$

Sign note: w and y must have the same sign in order to make $wx^2 y^3 z^4 = 2^{30}$ and $yz^2 w^3 = 2^{26}$ positive. These are the given equations where w and y have odd powers. Therefore, either w and y are both positive or they are both negative.

Similarly, x and z must have the same sign in order to make $xy^2 z^3 = 2^{14}$ and $zw^2 x^3 = 2^{30}$ positive. Either x and z are both positive or they are both negative.

Solution to Exercise #45

(A) $(1 + x)(1 - x) = 1 - x + x - x^2 = \boxed{1 - x^2}$

(B) $(1 + x + x^2)(1 - x) = 1 + x + x^2 - x - x^2 - x^3 = \boxed{1 - x^3}$

(C) $(1 + x + x^2 + x^3)(1 - x) = 1 + x + x^2 + x^3 - x - x^2 - x^3 - x^4 = \boxed{1 - x^4}$

(D) $(1 + x + x^2 + x^3 + x^4)(1 - x) = 1 + x + x^2 + x^3 + x^4 - x - x^2 - x^3 - x^4 - x^5$

$$= \boxed{1 - x^5}$$

(E) $(1 + x + x^2 + x^3 + \cdots + x^n)(1 - x) = 1 - x^{n+1}$

Divide both sides of the equation by $(1 - x)$. See the note in Part F.

$$1 + x + x^2 + x^3 + \cdots + x^n = \frac{1 - x^{n+1}}{1 - x} = \frac{P}{Q}$$

$$P = \boxed{1 - x^{n+1}} \quad , \quad Q = \boxed{1 - x}$$

(F) If $x = 1$, this causes the indeterminate form of zero divided by zero.

Solution to Exercise #46

$$(A) \sqrt{(x+c)^2 + y^2} + \sqrt{(x-c)^2 + y^2} = 2a$$

Square both sides. Apply the "foil" method: $(a+b)^2 = a^2 + 2ab + b^2$. Recall that $\sqrt{u}\sqrt{u} = u$ and $\sqrt{p}\sqrt{q} = \sqrt{pq}$.

$$(x+c)^2 + y^2 + 2\sqrt{(x+c)^2 + y^2}\sqrt{(x-c)^2 + y^2} + (x-c)^2 + y^2 = 4a^2$$

$$x^2 + 2cx + c^2 + y^2 + 2\sqrt{[(x+c)^2 + y^2][(x-c)^2 + y^2]} + x^2 - 2cx + c^2 + y^2$$
$$= 4a^2$$

$$2x^2 + 2c^2 + 2y^2 + 2\sqrt{(x+c)^2(x-c)^2 + (x+c)^2 y^2 + y^2(x-c)^2 + y^4} = 4a^2$$

$$x^2 + c^2 + y^2 + \sqrt{[(x+c)(x-c)]^2 + (x+c)^2 y^2 + y^2(x-c)^2 + y^4} = 2a^2$$

Recall that $(x+c)(x-c) = x^2 - c^2$ such that $[(x+c)(x-c)]^2 = (x^2 - c^2)^2$.

$$x^2 + c^2 + y^2 + \sqrt{(x^2 - c^2)^2 + (x^2 + 2cx + c^2)y^2 + (x^2 - 2cx + c^2)y^2 + y^4} = 2a^2$$

$$\sqrt{(x^2 - c^2)^2 + (x^2 + 2cx + c^2 + x^2 - 2cx + c^2)y^2 + y^4} = 2a^2 - x^2 - y^2 - c^2$$

$$\sqrt{(x^2 - c^2)^2 + (2x^2 + 2c^2)y^2 + y^4} = 2a^2 - x^2 - y^2 - c^2$$

$$\sqrt{x^4 - 2c^2 x^2 + c^4 + 2x^2 y^2 + 2c^2 y^2 + y^4} = 2a^2 - x^2 - y^2 - c^2$$

$$x^4 - 2c^2 x^2 + c^4 + 2x^2 y^2 + 2c^2 y^2 + y^4$$
$$= 4a^4 + x^4 + y^4 + c^4 - 4a^2 x^2 - 4a^2 y^2 - 4a^2 c^2 + 2x^2 y^2 + 2c^2 x^2 + 2c^2 y^2$$

$$-4c^2 x^2 = 4a^4 - 4a^2 x^2 - 4a^2 y^2 - 4a^2 c^2$$

$$c^2 x^2 = -a^4 + a^2 x^2 + a^2 y^2 + a^2 c^2$$

$$a^2 = b^2 + c^2$$

$$a^2 - b^2 = c^2$$

$$(a^2 - b^2)x^2 = -a^4 + a^2 x^2 + a^2 y^2 + a^2(a^2 - b^2)$$

$$a^2 x^2 - b^2 x^2 = -a^4 + a^2 x^2 + a^2 y^2 + a^4 - a^2 b^2$$

$$-b^2 x^2 = a^2 y^2 - a^2 b^2$$

$$a^2 b^2 = b^2 x^2 + a^2 y^2$$

$$\frac{a^2 b^2}{a^2 b^2} = \frac{b^2 x^2}{a^2 b^2} + \frac{a^2 y^2}{a^2 b^2}$$

$$\boxed{1 = \frac{x^2}{a^2} + \frac{y^2}{b^2}}$$

(B) $1 = \dfrac{5^2}{10^2} + \dfrac{y^2}{6^2}$

$1 = \dfrac{25}{100} + \dfrac{y^2}{36}$

$1 - \dfrac{25}{100} = 1 - \dfrac{1}{4} = \dfrac{3}{4} = \dfrac{y^2}{36}$

$\dfrac{3}{4}(36) = 27 = y^2$

$\pm\sqrt{27} = \pm\sqrt{(9)(3)} = \pm\sqrt{9}\sqrt{3}$

$= \boxed{\pm 3\sqrt{3}} = y$

$a^2 = b^2 + c^2$

$10^2 = 6^2 + c^2$

$100 = 36 + c^2$

$100 - 36 = 64 = c^2$

$\sqrt{64} = \boxed{8} = c$

(C) $\sqrt{(5+8)^2 + \left(\pm 3\sqrt{3}\right)^2}$

$+ \sqrt{(5-8)^2 + \left(\pm 3\sqrt{3}\right)^2} = 2(10)$

$\sqrt{13^2 + 3^2\left(\sqrt{3}\right)^2} + \sqrt{(-3)^2 + 3^2\left(\sqrt{3}\right)^2} = 20$

$\sqrt{169 + 9(3)} + \sqrt{9 + 9(3)} = 20$

$\sqrt{169 + 27} + \sqrt{36} = 20$

$\sqrt{196} + \sqrt{36} = 20$

$14 + 6 = 20 \quad \checkmark$

$a^2 = b^2 + c^2$

$10^2 = 6^2 + 8^2$

$100 = 36 + 64 \quad \checkmark$

$\dfrac{x^2}{a^2} + \dfrac{y^2}{b^2} = 1$

$\dfrac{5^2}{10^2} + \dfrac{\left(\pm 3\sqrt{3}\right)^2}{6^2} = 1$

$\dfrac{25}{100} + \dfrac{3^2\left(\sqrt{3}\right)^2}{36} = 1$

$\dfrac{1}{4} + \dfrac{9(3)}{36} = \dfrac{1}{4} + \dfrac{27}{36} = \dfrac{1}{4} + \dfrac{3}{4} = 1 \quad \checkmark$

Note: The given equations can be interpreted as the definition of an ellipse. The algebra involved in this problem derives the equation of an ellipse from this definition.

Solution to Exercise #47	
(A) $-\dfrac{ax}{6} > \dfrac{5}{3} - \dfrac{7}{4}$	Subtract $\dfrac{7}{4}$ from both sides.
$-\dfrac{ax}{6} > \dfrac{20}{12} - \dfrac{21}{12}$	Make a common denominator.
$-\dfrac{ax}{6} > -\dfrac{1}{12}$	Multiply both sides by 6.

$$-ax > -\frac{6}{12}$$	$\frac{6}{12}$ reduces to $\frac{1}{2}$.
$$-ax > -\frac{1}{2}$$	Multiply both sides by -1. This **reverses** the direction of the inequality. For example, compare $4 > 3$ and $-4 < -3$.
$$\boxed{ax < \frac{1}{2}}$$	
(B) $2x < \dfrac{1}{2}$	Divide both sides by 2.
$$\boxed{x < \frac{1}{4}}$$	
(C) $-2x < \dfrac{1}{2}$	Divide both sides by -2. This **reverses** the direction of the inequality.
$$\boxed{x > -\frac{1}{4}}$$	

(D) If a is positive, $x < \frac{1}{2a}$, but if a is negative, $x > \frac{1}{2a}$. When multiplying or dividing both sides of an inequality by a negative number, the inequality reverses direction. The answer for x depends on the sign of a.

(E) $x = 0.24$	$0.24 < 0.25$.
$$\frac{7}{4} - \frac{2(0.24)}{6} > \frac{5}{3}$$	$$\frac{7}{4} - \frac{2(0.24)}{6} = 1.75 - 0.08 = 1.67$$
$1.67 > 1.66666666...$ ✔	
$x = 0.27$	0.27 is not < 0.25.
$$\frac{7}{4} - \frac{2(0.27)}{6} = 1.66$$	$$\frac{7}{4} - \frac{2(0.27)}{6} = 1.75 - 0.09 = 1.66$$
1.66 is NOT $> 1.66666666...$ ✔	
(F) $x = -0.24$	$-0.24 > -0.25$ since -0.24 is not as negative as -0.25.
$$\frac{7}{4} - \frac{(-2)(-0.24)}{6} > \frac{5}{3}$$	
$1.67 > 1.66666666...$ ✔	$$\frac{7}{4} - \frac{(-2)(-0.24)}{6} = 1.75 - 0.08 = 1.67$$
$x = -0.27$	-0.27 is not > -0.25.
$$\frac{7}{4} - \frac{(-2)(-0.27)}{6} = 1.66$$	$$\frac{7}{4} - \frac{(-2)(-0.27)}{6} = 1.75 - 0.09 = 1.66$$
1.66 is NOT $> 1.66666666...$ ✔	

Note: $\frac{5}{3} = 1.66666666\ldots$ is a repeating decimal; the digit 6 repeats forever. This is usually written with an overbar: $1.\overline{6}$.

Solution to Exercise #48

(A) $y = -4x - 9$

$5(-4x - 9) + 27 = 12x - 8x^2$

$-20x - 45 + 27 = 12x - 8x^2$

$8x^2 - 32x - 18 = 0$

$4x^2 - 16x - 9 = 0$

$(2x + 1)(2x - 9) = 0$

$2x + 1 = 0$ or $2x - 9 = 0$

$2x = -1$ or $2x = 9$

$x = \boxed{-\dfrac{1}{2}} = \boxed{-0.5}$ or $x = \boxed{\dfrac{9}{2}} = \boxed{4.5}$

$y = -4x - 9 = -4\left(-\dfrac{1}{2}\right) - 9$

$= 2 - 9 = \boxed{-7}$

$y = -4x - 9 = -4\left(\dfrac{9}{2}\right) - 9$

$= -18 - 9 = \boxed{-27}$

(B) $5(-7) + 27 = -35 + 27 = -8$

$12\left(-\dfrac{1}{2}\right) - 8\left(-\dfrac{1}{2}\right)^2 = -6 - 8\left(\dfrac{1}{4}\right)$

$= -6 - 2 = -8$ ✓

$-7 + 9 = 2 = -4\left(-\dfrac{1}{2}\right) = 2$ ✓

$5(-27) + 27 = -135 + 27 = -108$

$12\left(\dfrac{9}{2}\right) - 8\left(\dfrac{9}{2}\right)^2 = 54 - 8\left(\dfrac{81}{4}\right)$

$= 54 - 162 = -108$ ✓

$-27 + 9 = -18 = -4\left(\dfrac{9}{2}\right) = -18$ ✓

Solution to Exercise #49

$16x^7y^2 + 9\sqrt{(x^7 - 1)x^2y^4} = 25[x^6(y^3 + 1)]^{2/3}$

$16x^7y^2 + 9\sqrt{x^7 - 1}\sqrt{x^2y^4} = 25(x^6)^{2/3}(y^3 + 1)^{2/3}$

$16x^7y^2 + 9xy^2\sqrt{x^7 - 1} = 25x^4(y^3 + 1)^{2/3}$

$\dfrac{16x^7y^2}{x^4y^2} + \dfrac{9xy^2\sqrt{x^7 - 1}}{x^4y^2} = \dfrac{25x^4(y^3 + 1)^{2/3}}{x^4y^2}$

$\boxed{16x^3 + \dfrac{9\sqrt{x^7 - 1}}{x^3} = \dfrac{25(y^3 + 1)^{2/3}}{y^2}}$

Factor x^2y^4 and x^6.

Apply $(t^a)^b = t^{ab}$, $\sqrt{pq} = \sqrt{p}\sqrt{q}$, and $(cd)^e = c^e d^e$.

Divide both sides by x^4y^2.

Note: Separation of variables is a useful technique in advanced mathematics (especially, differential equations). Of course, knowledge of advanced math is NOT needed to solve this algebra problem.

Solution to Exercise #50

(A) $u^2 + x^2 = w^2 + 2y^2 + z^2$

$w = r - t$, $y^2 = n^2 - t^2$

$z = r + t$

$u^2 + x^2 = (r-t)^2 + 2(n^2 - t^2)$
$\qquad + (r+t)^2$

$u^2 + x^2 = r^2 - 2rt + t^2 + 2n^2 - 2t^2$
$\qquad + r^2 + 2rt + t^2$

$\boxed{u^2 + x^2 = 2r^2 + 2n^2}$

(B) $q^2 + y^2 = n^2$

$5^2 + y^2 = 13^2$

$25 + y^2 = 169$

$y^2 = 169 - 25 = 144$

$y = \sqrt{144} = \boxed{12}$

$q^2 + y^2 = t^2 + y^2$

$t = q = \boxed{5}$

$z = r + t = 29 + 5 = \boxed{34}$

$w = r - t = 29 - 5 = \boxed{24}$

$w = p - q$

$24 = p - 5$

$\boxed{29} = p$

$u^2 = w^2 + y^2 = 24^2 + 12^2$

$= 576 + 144 = 720$

$u = \sqrt{720} = \sqrt{(144)(5)} = \sqrt{144}\sqrt{5}$

$= \boxed{12\sqrt{5}}$

$x^2 = y^2 + z^2 = 12^2 + 34^2$

$= 144 + 1156 = 1300$

$x = \sqrt{1300} = \sqrt{(100)(13)}$

$= \sqrt{100}\sqrt{13} = \boxed{10\sqrt{13}}$

(C) $u^2 = w^2 + y^2$

$\left(12\sqrt{5}\right)^2 = 24^2 + 12^2$

$144(5) = 720 = 576 + 144$ ✓

$x^2 = y^2 + z^2$

$\left(10\sqrt{13}\right)^2 = 12^2 + 34^2$

$100(13) = 1300 = 144 + 1156$ ✓

$q^2 + y^2 = t^2 + y^2 = n^2$

$5^2 + 12^2 = 5^2 + 12^2 = 13^2$

$25 + 144 = 169$ ✓

$w = p - q = r - t$

$24 = 29 - 5 = 29 - 5$ ✓

$z = r + t$

$34 = 29 + 5$ ✓

$u^2 + x^2 = 2r^2 + 2n^2$

$\left(12\sqrt{5}\right)^2 + \left(10\sqrt{13}\right)^2 = 2(29)^2 + 2(13)^2$

$144(5) + 100(13) = 2(841) + 2(169)$

$720 + 1300 = 2020 = 1682 + 338$ ✓

Note: The algebra in this problem corresponds to a theorem regarding parallelograms. The sum of the squares of the diagonals $(u^2 + x^2)$ equals the sum of the squares of the sides $(p^2 + n^2 + r^2 + n^2 = 2r^2 + 2n^2)$.

Solution to Exercise #51

(A) $\left[\dfrac{x}{x+1}\left(\dfrac{x}{x}\right)-\dfrac{x-1}{x}\left(\dfrac{x+1}{x+1}\right)\right]^{-1}-\left[\dfrac{x}{x-1}\left(\dfrac{x}{x}\right)-\dfrac{x+1}{x}\left(\dfrac{x-1}{x-1}\right)\right]^{-1}$

$=\left[\dfrac{x^2}{(x+1)x}-\dfrac{x^2+x-x-1}{(x+1)x}\right]^{-1}-\left[\dfrac{x^2}{(x-1)x}-\dfrac{x^2-x+x-1}{(x-1)x}\right]^{-1}$

$=\left(\dfrac{x^2}{x^2+x}-\dfrac{x^2-1}{x^2+x}\right)^{-1}-\left(\dfrac{x^2}{x^2-x}-\dfrac{x^2-1}{x^2-x}\right)^{-1}$

$=\left[\dfrac{x^2-(x^2-1)}{x^2+x}\right]^{-1}-\left[\dfrac{x^2-(x^2-1)}{x^2-x}\right]^{-1}=\left(\dfrac{x^2-x^2+1}{x^2+x}\right)^{-1}-\left(\dfrac{x^2-x^2+1}{x^2-x}\right)^{-1}$

$=\left(\dfrac{1}{x^2+x}\right)^{-1}-\left(\dfrac{1}{x^2-x}\right)^{-1}=x^2+x-(x^2-x)=x^2+x-x^2+x=\boxed{2x}$

(B) $2x=2(4)=\boxed{8}$

$\left(\dfrac{4}{4+1}-\dfrac{4-1}{4}\right)^{-1}-\left(\dfrac{4}{4-1}-\dfrac{4+1}{4}\right)^{-1}=\left(\dfrac{4}{5}-\dfrac{3}{4}\right)^{-1}-\left(\dfrac{4}{3}-\dfrac{5}{4}\right)^{-1}$

$=\left(\dfrac{16}{20}-\dfrac{15}{20}\right)^{-1}-\left(\dfrac{16}{12}-\dfrac{15}{12}\right)^{-1}=\left(\dfrac{1}{20}\right)^{-1}-\left(\dfrac{1}{12}\right)^{-1}=20-12=\boxed{8}$ ✓

(C) $x=-1, x=0,$ or $x=1$ would cause division by zero.

Solution to Exercise #52

$aw+bx=ay+bz$	Subtract ay and bx from both sides.
$aw-ay=bz-bx$	Factor out a and b.
$a(w-y)=b(z-x)$	Divide both sides by $w-y$.
$\dfrac{a}{b}=\dfrac{z-x}{w-y}$	
$aw^2+bx^2=ay^2+bz^2$	Subtract ay^2 and bx^2 from both sides.
$aw^2-ay^2=bz^2-bx^2$	Factor out a and b.
$a(w^2-y^2)=b(z^2-x^2)$	Divide both sides by w^2-y^2.
$\dfrac{a}{b}=\dfrac{z^2-x^2}{w^2-y^2}=\dfrac{(z+x)(z-x)}{(w+y)(w-y)}$	Recall that $(p+q)(p-q)=p^2-q^2$.
$\dfrac{a}{b}=\dfrac{(z+x)(z-x)}{(w+y)(w-y)}=\dfrac{z-x}{w-y}$	Set $\dfrac{a}{b}=\dfrac{z-x}{w-y}$ equal to $\dfrac{a}{b}=\dfrac{(z+x)(z-x)}{(w+y)(w-y)}$.

$\dfrac{z + x}{w + y} = 1$	Let $F = \dfrac{z+x}{w+y}$ and $G = \dfrac{z-x}{w-y}$. In the
$\boxed{z + x = w + y}$	previous equation, $FG = G$. This
Note: The solution to this problem is	requires that either $F = 1$ or $G = 0$.
featured on the front cover of this book.	Based on the wording of the problem, in
	this case only $F = 1 = \dfrac{z+x}{w+y}$ applies.

Note: Algebra similar to this problem arises in physics in the context of elastic one-dimensional collisions between two objects.

Solution to Exercise #53	
(A) $\dfrac{6 - 2x}{3x + 9} = \dfrac{x - 10}{4x - 6}$	(B) $\dfrac{6 - 2\left(-\dfrac{9}{11}\right)}{3\left(-\dfrac{9}{11}\right) + 9} = \dfrac{\dfrac{66}{11} + \dfrac{18}{11}}{-\dfrac{27}{11} + \dfrac{99}{11}} = \dfrac{\dfrac{84}{11}}{\dfrac{72}{11}}$
Cross multiply: $\dfrac{a}{b} = \dfrac{c}{d}$ becomes $ad = bc$.	$= \dfrac{84}{11} \div \dfrac{72}{11} = \dfrac{84}{11} \times \dfrac{11}{72} = \dfrac{84}{72} = \dfrac{7}{6}$
$(6 - 2x)(4x - 6) = (x - 10)(3x + 9)$	
$24x - 36 - 8x^2 + 12x$	$\dfrac{-\dfrac{9}{11} - 10}{4\left(-\dfrac{9}{11}\right) - 6} = \dfrac{-\dfrac{9}{11} - \dfrac{110}{11}}{-\dfrac{36}{11} - \dfrac{66}{11}} = \dfrac{-\dfrac{119}{11}}{-\dfrac{102}{11}}$
$= 3x^2 + 9x - 30x - 90$	
$36x - 36 - 8x^2 = 3x^2 - 21x - 90$	$= \dfrac{119}{11} \div \dfrac{102}{11} = \dfrac{119}{11} \times \dfrac{11}{102}$
$0 = 11x^2 - 57x - 54$	$= \dfrac{119}{102} = \dfrac{7}{6}$ ✓
$(11x + 9)(x - 6) = 0$	$\dfrac{6 - 2(6)}{3(6) + 9} = \dfrac{6 - 12}{18 + 9} = \dfrac{-6}{27} = -\dfrac{2}{9}$
$11x + 9 = 0$ or $x - 6 = 0$	
$11x = -9$ or $x = 6$	$\dfrac{x - 10}{4x - 6} = \dfrac{6 - 10}{4(6) - 6} = \dfrac{-4}{24 - 6}$
$x = \boxed{-\dfrac{9}{11}}$ or $x = \boxed{6}$	$= -\dfrac{4}{18} = -\dfrac{2}{9}$ ✓

Solution to Exercise #54	
(A) $z = \sqrt{u^2 + \left(wp - \dfrac{1}{wq}\right)^2}$	(D) $x = wp = 18w$
	$y = \dfrac{1}{wq} = \dfrac{1}{w/450} = \dfrac{450}{w}$

$$\left(wp - \frac{1}{wq}\right)^2 \geq 0$$

$\left(wp - \frac{1}{wq}\right)^2 \geq 0$ because the square of any (real) number can not be negative.

z is minimum when $\left(wp - \frac{1}{wq}\right)^2$ is 0.

$$z_{min} = \sqrt{u^2 + 0} = \boxed{u}$$

For a given value of u, the minimum possible value of z equals u: $z \geq u$.

$$(B)\ wp - \frac{1}{wq} = 0$$

$$wp = \frac{1}{wq}$$

$$w^2 = \frac{1}{pq}$$

$$\boxed{w = \frac{1}{\sqrt{pq}}} = \frac{1}{\sqrt{pq}}\frac{\sqrt{pq}}{\sqrt{pq}} = \boxed{\frac{\sqrt{pq}}{pq}}$$

$$(C)\ z_{min} = u = \boxed{180}$$

$$w = \frac{1}{\sqrt{pq}} = \frac{1}{\sqrt{(18)\left(\frac{1}{450}\right)}} = \frac{1}{\sqrt{\frac{18}{450}}}$$

$$= \frac{1}{\sqrt{\frac{9}{225}}} = \frac{1}{\frac{\sqrt{9}}{\sqrt{225}}} = \frac{1}{3/15}$$

$$= \frac{1}{1/5} = \boxed{5}$$

$$z = \sqrt{u^2 + (x - y)^2}$$

$$225 = \sqrt{180^2 + \left(18w - \frac{450}{w}\right)^2}$$

$$225^2 = 32{,}400 + \left(18w - \frac{450}{w}\right)^2$$

$$50{,}625 - 32{,}400 = 18{,}225$$

$$= \left(18w - \frac{450}{w}\right)^2$$

$$\pm\sqrt{18{,}225} = \pm 135 = 18w - \frac{450}{w}$$

$$\pm 135w = 18w^2 - 450$$

$$\pm 15w = 2w^2 - 50$$

$$0 = 2w^2 \mp 15w - 50$$

$$w = \frac{\pm 15 \pm \sqrt{(\mp 15)^2 - 4(2)(-50)}}{2(2)}$$

$$= \frac{\pm 15 \pm \sqrt{225 + 400}}{4}$$

$$w = \frac{\pm 15 \pm \sqrt{625}}{4} = \frac{\pm 15 \pm 25}{4}$$

$$w = \frac{-15 + 25}{4} \quad \text{or} \quad w = \frac{15 + 25}{4}$$

$$w = \frac{10}{4} \quad \text{or} \quad w = \frac{40}{4}$$

$$w = \boxed{\frac{5}{2}} = \boxed{2.5} \quad \text{or} \quad w = \boxed{10}$$

Note: This problem is similar to RLC circuits. In RLC circuits, the equations look like $Z = \sqrt{R^2 + (X_L - X_C)^2}$, $X_L = \omega L$, and $X_C = \frac{1}{\omega C}$. The value of ω that minimizes Z is called the resonance frequency: $\omega_0 = \frac{1}{\sqrt{LC}}$. The minimum value of Z equals R.

Solution to Exercise #55

$$r^2 = z^2 \left(\frac{p^2}{x^2} + \frac{q^2}{y^2} \right)$$	Plug in $z = \frac{ax}{y^2}$.
$$r^2 = \left(\frac{ax}{y^2} \right)^2 \left(\frac{p^2}{x^2} + \frac{q^2}{y^2} \right)$$	Apply the rule $(b^c)^d = b^{cd}$.
$$r^2 = \frac{a^2 x^2}{y^4} \left(\frac{p^2}{x^2} + \frac{q^2}{y^2} \right)$$	Distribute x^2.
$$r^2 = \frac{a^2}{y^4} \left(\frac{p^2 x^2}{x^2} + \frac{q^2 x^2}{y^2} \right)$$	$$\frac{x^2}{x^2} = 1$$
$$r^2 = \frac{a^2}{y^4} \left(p^2 + \frac{q^2 x^2}{y^2} \right)$$	Factor out p^2. Check by distributing:
$$r^2 = \frac{a^2 p^2}{y^4} \left(1 + \frac{q^2 x^2}{p^2 y^2} \right)$$	$$p^2 \left(1 + \frac{q^2 x^2}{p^2 y^2} \right) = p^2 + \frac{q^2 x^2}{y^2}$$
$$r^2 = \frac{a^2 p^2}{y^4} \left[1 + \left(\frac{qx}{py} \right)^2 \right]$$	Square root both sides. Consider both positive and negative roots. The problem did NOT restrict the sign of r.
$$\boxed{r = \pm \frac{ap}{y^2} \sqrt{1 + \left(\frac{qx}{py} \right)^2}}$$	Note that $(-3)^2 = 9$ and $3^2 = 9$, for example.

Solution to Exercise #56

(A) $x = yz$	(B) $x = yz$
$$y = \left(z + \frac{3}{2} \right) x$$	$$1 = \left(-\frac{1}{2} \right)(-2) \quad \checkmark$$
$$y = \left(z + \frac{3}{2} \right) yz$$	$$-4 = (2)(-2) \quad \checkmark$$
$$1 = \left(z + \frac{3}{2} \right) z$$	$$\frac{-3 \pm \sqrt{10}}{2} = (-3 \pm \sqrt{10}) \left(\frac{1}{2} \right) \quad \checkmark$$
$$1 = z^2 + \frac{3z}{2}$$	$$y = \left(z + \frac{3}{2} \right) x$$
$$0 = z^2 + \frac{3z}{2} - 1$$	$$-\frac{1}{2} = \left(-2 + \frac{3}{2} \right)(1) = \left(-\frac{1}{2} \right)(1) \quad \checkmark$$
$$0 = 2z^2 + 3z - 2$$	$$2 = \left(-2 + \frac{3}{2} \right)(-4) = \left(-\frac{1}{2} \right)(-4) \quad \checkmark$$

$$z = \frac{-3 \pm \sqrt{3^2 - 4(2)(-2)}}{2(2)}$$

$$z = \frac{-3 \pm \sqrt{9 + 16}}{4} = \frac{-3 \pm \sqrt{25}}{4} = \frac{-3 \pm 5}{4}$$

$$z = \frac{-3 - 5}{4} \quad \text{or} \quad z = \frac{-3 + 5}{4}$$

$$z = \frac{-8}{4} \quad \text{or} \quad z = \frac{2}{4}$$

$$z = \boxed{-2} \quad \text{or} \quad z = \boxed{\frac{1}{2}} = \boxed{0.5}$$

$$z = (x + 3)y$$
$$z = (yz + 3)y$$
$$z = y^2 z + 3y$$

$$-2 = -2y^2 + 3y \quad \text{or} \quad \frac{1}{2} = \frac{y^2}{2} + 3y$$

$$2y^2 - 3y - 2 = 0 \quad \text{or} \quad 1 = y^2 + 6y$$

$$2y^2 - 3y - 2 = 0 \quad \text{or} \quad 0 = y^2 + 6y - 1$$

$$y = \frac{-(-3) \pm \sqrt{(-3)^2 - 4(2)(-2)}}{2(2)}$$

$$\text{or} \quad y = \frac{-6 \pm \sqrt{6^2 - 4(1)(-1)}}{2(1)}$$

$$y = \frac{3 \pm \sqrt{9 + 16}}{4} \quad \text{or} \quad y = \frac{-6 \pm \sqrt{36 + 4}}{2}$$

$$y = \frac{3 \pm \sqrt{25}}{4} = \frac{-3 \pm 5}{4}$$

$$\text{or} \quad y = \frac{-6 \pm \sqrt{40}}{2} = \frac{-6 \pm 2\sqrt{10}}{2}$$

$$= -3 \pm \sqrt{10}$$

$$y = \frac{3 - 5}{4} \quad , \quad y = \frac{3 + 5}{4}$$

$$\text{or} \quad y = -3 \pm \sqrt{10}$$

$$-3 \pm \sqrt{10} = \left(\frac{1}{2} + \frac{3}{2}\right)\left(\frac{-3 \pm \sqrt{10}}{2}\right)$$

$$= (2)\left(\frac{-3 \pm \sqrt{10}}{2}\right) \quad \checkmark$$

$$z = (x + 3)y$$

$$-2 = (1 + 3)\left(-\frac{1}{2}\right) = (4)\left(-\frac{1}{2}\right) \quad \checkmark$$

$$-2 = (-4 + 3)(2) = (-1)(2) \quad \checkmark$$

$$\frac{1}{2} = \left(\frac{-3 \pm \sqrt{10}}{2} + 3\right)(-3 \pm \sqrt{10})$$

$$= \left(\frac{-3 \pm \sqrt{10}}{2} + \frac{6}{2}\right)(-3 \pm \sqrt{10})$$

$$\frac{1}{2} = \left(\frac{3 \pm \sqrt{10}}{2}\right)(-3 \pm \sqrt{10})$$

$$= \frac{10 - 9}{2} = \frac{1}{2} \quad \checkmark$$

Note:

$$\left(3 \pm \sqrt{10}\right)\left(-3 \pm \sqrt{10}\right)$$
$$= 3(-3) \pm 3\sqrt{10} \mp 3\sqrt{10} + \sqrt{10}\sqrt{10}$$
$$= -9 + 10 = 10 - 9 = 1$$

$$y = \frac{-2}{4} \quad , \quad y = \frac{8}{4} \quad \text{or} \quad y = -3 \pm \sqrt{10}$$

$$y = \boxed{-\frac{1}{2}} = \boxed{-0.5} \quad \text{or} \quad y = \boxed{2}$$

$$\text{or} \quad y = \boxed{-3 \pm \sqrt{10}}$$

$$x = yz$$

$$x = \left(-\frac{1}{2}\right)(-2) = \boxed{1}$$

$$\text{or} \quad x = (2)(-2) = \boxed{-4}$$

$$\text{or} \quad x = \left(-3 \pm \sqrt{10}\right)\left(\frac{1}{2}\right) = \boxed{\frac{-3 \pm \sqrt{10}}{2}}$$

Either $x = 1$, $y = -\frac{1}{2}$, and $z = -2$,

or $x = -4$, $y = 2$, and $z = -2$,

or $x = \frac{-3 \pm \sqrt{10}}{2}$, $y = -3 \pm \sqrt{10}$, and $z = \frac{1}{2}$.

Solution to Exercise #57

$$z = \frac{4x^3\sqrt{x^2 + 4} - x^5(x^2 + 4)^{-1/2}}{x^2 + 4}$$

$$z = \frac{4x^3(x^2 + 4)^{1/2} - x^5(x^2 + 4)^{-1/2}}{x^2 + 4}$$

$$z = \frac{(x^2 + 4)^{-1/2}[4x^3(x^2 + 4) - x^5]}{x^2 + 4}$$

$$z = \frac{4x^3(x^2 + 4) - x^5}{(x^2 + 4)^{3/2}} = \frac{4x^5 + 16x^3 - x^5}{(x^2 + 4)^{3/2}}$$

$$z = \boxed{\frac{3x^5 + 16x^3}{(x^2 + 4)^{3/2}}} = \frac{P}{Q^k}$$

$$k = \boxed{\frac{3}{2}} = \boxed{1.5} \quad , \quad P = \boxed{3x^5 + 16x^3}$$

$$Q = \boxed{x^2 + 4}$$

Recall that $u^{1/2} = \sqrt{u}$ since $u^{1/2}u^{1/2} = u^1 = u$ and $\sqrt{u}\sqrt{u} = u$.

Factor out $(x^2 + 4)^{-1/2}$. Check by distributing:

$$(x^2 + 4)^{-1/2}[4x^3(x^2 + 4) - x^5]$$
$$= 4x^3(x^2 + 4)^{1/2} - x^5(x^2 + 4)^{-1/2}$$

because $u^{-1/2}u = u^{-1/2+1} = u^{1/2}$.

Note that $\frac{u^{-1/2}}{u} = u^{-1/2-1} = u^{-3/2} = \frac{1}{u^{3/2}}$.

Here, $u = x^2 + 4$.

Solution to Exercise #58

(A) $x + \dfrac{1}{y} = 2$

$$x = 2 - \frac{1}{y} = \frac{2y}{y} - \frac{1}{y} = \frac{2y - 1}{y}$$

Take the reciprocal of both sides.

$$\frac{1}{x} = \frac{y}{2y - 1}$$

$$z + \frac{1}{x} = 5$$

$$z + \frac{y}{2y - 1} = 5$$

$$z = 5 - \frac{y}{2y - 1} = \frac{5(2y - 1)}{2y - 1} - \frac{y}{2y - 1}$$

$$= \frac{10y - 5}{2y - 1} - \frac{y}{2y - 1}$$

$$= \frac{9y - 5}{2y - 1}$$

$$\frac{1}{z} = \frac{2y - 1}{9y - 5}$$

$$y + \frac{1}{z} = 1$$

$$y + \frac{2y - 1}{9y - 5} = 1$$

$$y(9y - 5) + 2y - 1 = 9y - 5$$

$$9y^2 - 5y + 2y - 1 = 9y - 5$$

$$9y^2 - 3y - 1 = 9y - 5$$

$$9y^2 - 12y + 4 = 0$$

$$(3y - 2)(3y - 2) = 0$$

$$3y - 2 = 0$$

$$3y = 2$$

$$y = \boxed{\frac{2}{3}}$$

(B) $x + \dfrac{1}{y} = 2$

$$\frac{1}{2} + \frac{1}{2/3} = \frac{1}{2} + \frac{3}{2} = \frac{4}{2} = 2 \quad \checkmark$$

$$y + \frac{1}{z} = 1$$

$$\frac{2}{3} + \frac{1}{3} = \frac{3}{3} = 1 \quad \checkmark$$

$$z + \frac{1}{x} = 5$$

$$3 + \frac{1}{1/2} = 3 + 2 = 5 \quad \checkmark$$

$$x + \frac{1}{y} = 2$$

$$x + \frac{1}{2/3} = 2$$

$$x + \frac{3}{2} = 2$$

$$x = 2 - \frac{3}{2} = \frac{4}{2} - \frac{3}{2} = \boxed{\frac{1}{2}}$$

$$z + \frac{1}{x} = 5$$

$$z + \frac{1}{1/2} = 5$$

$$z + 2 = 5$$

$$z = 5 - 2 = \boxed{3}$$

Solution to Exercise #59

$$k\frac{a}{x^2} = w \quad , \quad w = \frac{y^2}{x}$$

Apply the transitive property.

$$k\frac{a}{x^2} = \frac{y^2}{x}$$

Multiply both sides by x.

$$k\frac{a}{x} = y^2$$

$$y = \frac{cx}{z}$$

$$k\frac{a}{x} = \left(\frac{cx}{z}\right)^2$$

Cross multiply.

$$k\frac{a}{x} = \frac{c^2 x^2}{z^2}$$

$$\frac{q}{r} = \frac{s}{t}$$

$$kaz^2 = c^2 x^3$$

$$qt = rs$$

$$z^2 = \frac{c^2 x^3}{ka}$$

Square root both sides of the equation.

$$\boxed{z = c\sqrt{\frac{x^3}{ka}}} = \frac{c\sqrt{x^3}\sqrt{ka}}{\sqrt{ka}\sqrt{ka}} = \boxed{\frac{c\sqrt{kax^3}}{ka}}$$

Since the problem states that $z > 0$, only the positive root applies.

Note: The equations in this problem are similar in structure to the equations of a satellite in a circular orbit. For a satellite in a circular orbit, $\frac{G m_p m_s}{R^2} = m a_c$, $a_c = \frac{v^2}{R}$, and $v = \frac{2\pi R}{T}$, which leads to a period of $T = 2\pi \sqrt{\frac{R^3}{G m_p}}$. This is consistent with Kepler's third law of planetary motion, which states that the square of the orbital period is proportional to the cube of the semimajor axis. (This problem works out the algebra for a circular orbit with radius R. The semimajor axis described in Kepler's third law refers to the more general case of an elliptical orbit.)

Solution to Exercise #60

$$z = \sqrt{7 + \sqrt{5}} - \sqrt{7 - \sqrt{5}}$$

Square both sides of the equation. Apply the "foil" method:

$$(p - q)^2 = p^2 - 2pq + q^2$$

$$z^2 = \sqrt{7 + \sqrt{5}}\sqrt{7 + \sqrt{5}} - \sqrt{7 + \sqrt{5}}\sqrt{7 - \sqrt{5}} - \sqrt{7 - \sqrt{5}}\sqrt{7 + \sqrt{5}} + \sqrt{7 - \sqrt{5}}\sqrt{7 - \sqrt{5}}$$

Apply the rule $\sqrt{t}\sqrt{t} = t$.

$$z^2 = 7 + \sqrt{5} - 2\sqrt{7 + \sqrt{5}}\sqrt{7 - \sqrt{5}} + 7 - \sqrt{5}$$

Apply the rule $\sqrt{p}\sqrt{q} = \sqrt{pq}$.

$$z^2 = 14 - 2\sqrt{(7 + \sqrt{5})(7 - \sqrt{5})}$$

$$z^2 = 14 - 2\sqrt{49 - 7\sqrt{5} + 7\sqrt{5} - 5}$$

$$z^2 = 14 - 2\sqrt{44} = 14 - 2\sqrt{(4)(11)} = 14 - 2\sqrt{4}\sqrt{11} = 14 - 2(2)\sqrt{11} = 14 - 4\sqrt{11}$$

$$z^2 = x + w\sqrt{y}$$

$$x = \boxed{14} \quad , \quad w = \boxed{-4} \quad , \quad y = \boxed{11}$$

The solution below is a viable alternative, but is not the simplest form. These are equivalent solutions because $2\sqrt{44} = 4\sqrt{11}$ (as shown in the steps of math above).

$$x = \boxed{14} \quad , \quad w = \boxed{-2} \quad , \quad y = \boxed{44}$$

Check the answer using a calculator: $z \approx 0.856446635$ and $z^2 \approx 0.733500839$.

Solution to Exercise #61

(A) $8\sqrt{x} - \dfrac{4\sqrt{x}}{\sqrt{x}} = \dfrac{\sqrt{x}\sqrt{x}}{x}$	Multiply both sides by $\sqrt{x}$.
$8\sqrt{x} - 4 = \dfrac{x}{x}$	Note that $\sqrt{x}\sqrt{x} = x$ and $\dfrac{\sqrt{x}}{\sqrt{x}} = 1$.
$8\sqrt{x} - 4 = 1$	Add 4 to both sides.
$8\sqrt{x} = 5$	Divide both sides by 8.
$\sqrt{x} = \dfrac{5}{8}$	Square both sides.
$\left(\sqrt{x}\right)^2 = \left(\dfrac{5}{8}\right)^2$	Recall that $\left(\dfrac{a}{b}\right)^2 = \dfrac{a^2}{b^2}$.
$x = \boxed{\dfrac{25}{64}} = \boxed{0.390625}$	
(B) $8 - \dfrac{4}{\sqrt{25/64}} = \dfrac{\sqrt{25/64}}{25/64}$	$\sqrt{\dfrac{25}{64}} = \dfrac{\sqrt{25}}{\sqrt{64}} = \dfrac{5}{8}$
$8 - \dfrac{4}{5/8} = \dfrac{5/8}{25/64}$	$4 \div \dfrac{5}{8} = 4 \times \dfrac{8}{5}$ and $\dfrac{5}{8} \div \dfrac{25}{64} = \dfrac{5}{8} \times \dfrac{64}{25}$.
$8 - 4\left(\dfrac{8}{5}\right) = \dfrac{5}{8}\left(\dfrac{64}{25}\right)$	
$\dfrac{40}{5} - \dfrac{32}{5} = \dfrac{8}{5}$ ✓	

Solution to Exercise #62

(A) $x = (4 + w)y$	
$x = (4 - w)z$	Multiply the two equations. Recall that
$x^2 = (16 - w^2)yz$	$(4 + w)(4 - w) = 16 - w^2$
$yz = 156$	Replace yz with 156.
$x^2 = (16 - w^2)156$	Distribute: $(a - b)c = ac - bc$.
$= (16)(156) - 156w^2$	
$5x = 78w$	This is one of the given equations.
$x = \dfrac{78w}{5}$	Divide both sides by 5.

$$x^2 = \frac{(78)(78)w^2}{25} = (16)(156) - 156w^2$$

$$(78)(78)w^2 = (400)(156) - (25)(156)w^2$$

$$78w^2 = (400)(2) - (25)(2)w^2$$

$$39w^2 = 400 - 25w^2$$

$$64w^2 = 400$$

$$w^2 = \frac{400}{64}$$

$$w = \pm\sqrt{\frac{400}{64}} = \pm\frac{\sqrt{400}}{\sqrt{64}}$$

$$= \pm\frac{20}{8} = \boxed{\pm\frac{5}{2}} = \boxed{\pm 2.5}$$

$$5x = 78w = 78\left(\pm\frac{5}{2}\right) = \pm 195$$

$$x = \pm\frac{195}{5} = \boxed{\pm 39}$$

$$x = (4 - w)z$$

$$39 = \left(4 - \frac{5}{2}\right)z$$

$$\text{or} \quad -39 = \left[4 - \left(-\frac{5}{2}\right)\right]z$$

$$39 = \left(\frac{8}{2} - \frac{5}{2}\right)z \quad \text{or} \quad -39 = \left(\frac{8}{2} + \frac{5}{2}\right)z$$

$$39 = \frac{3z}{2} \quad \text{or} \quad -39 = \frac{13z}{2}$$

$$\boxed{26} = z \quad \text{or} \quad -\frac{390}{13} = \boxed{-6} = z$$

$$yz = 156$$

$$y = \frac{156}{z}$$

$$y = \frac{156}{26} \quad \text{or} \quad y = \frac{156}{-6}$$

$$y = \boxed{6} \quad \text{or} \quad y = \boxed{-26}$$

Square both sides of the equation. Combine $x^2 = \frac{(78)(78)w^2}{25}$ with the equation $x^2 = (16)(156) - 156w^2$ from earlier. Multiply both sides by 25. Divide both sides by 78. Add $25w^2$ to both sides. Divide both sides by 64. Square root both sides. Allow for both positive and negative roots. Note, for example, that $(-8)^2 = 64$ and $8^2 = 64$.

$z = 26$ corresponds to $w = \frac{5}{2}$ and $x = 39$, while $z = -6$ corresponds to $w = -\frac{5}{2}$ and $x = -39$.

$y = 6$ corresponds to $z = 26$, while $y = -26$ corresponds to $z = -6$.

$$(B)\ x = (4 + w)y$$
$$39 = (4 + 2.5)(6) = (6.5)(6) \quad \checkmark$$
$$-39 = (4 - 2.5)(-26)$$
$$= (1.5)(-26) \quad \checkmark$$
$$x = (4 - w)z$$
$$39 = (4 - 2.5)(26) = (1.5)(26) \quad \checkmark$$
$$-39 = [4 - (-2.5)](-6)$$
$$= (6.5)(-6) \quad \checkmark$$
$$yz = 156$$
$$(6)(26) = 156 \quad \checkmark$$
$$(-26)(-6) = 156 \quad \checkmark$$
$$5x = 78w$$
$$5(\pm 39) = 78(\pm 2.5) = \pm 195 \quad \checkmark$$

Solution to Exercise #63

$$(A)\ \sqrt{x^2 + (y - p)^2} = |y + p|$$
$$x^2 + (y - p)^2 = y^2 + 2py + p^2$$
$$x^2 + y^2 - 2py + p^2 = y^2 + 2py + p^2$$
$$x^2 = 4py$$
$$\frac{x^2}{4p} = y$$
$$a = \frac{1}{4p}$$
$$\boxed{ax^2 = y}$$
$$(B)\ y = ax^2 = \frac{1}{12}(18)^2 = \frac{324}{12} = \boxed{27}$$
$$a = \frac{1}{12} = \frac{1}{4p}$$
$$4p = 12$$
$$p = \frac{12}{4} = \boxed{3}$$

Square both sides of the equation. $(y + p)^2$ is positive regardless of the sign of $y + p$, so the absolute values are no longer needed.

$$(y + p)^2 = y^2 + 2py + p^2$$
$$(y - p)^2 = y^2 - 2py + p^2$$

Note that:

$$ax^2 = \left(\frac{1}{4p}\right)x^2 = \frac{x^2}{4p}$$

Note: The given equations can be interpreted as the definition of a parabola. The algebra involved in this problem derives the equation of a parabola from this definition.

$$\text{(C) } \sqrt{x^2 + (y-p)^2} = |y + p|$$

$$\sqrt{18^2 + (27-3)^2} = |27 + 3|$$

$$\sqrt{324 + 24^2} = \sqrt{324 + 576} = \sqrt{900} = 30 \quad \checkmark$$

$$a = \frac{1}{4p}$$

$$\frac{1}{12} = \frac{1}{4(3)} \quad \checkmark$$

$$y = ax^2$$

$$27 = \frac{1}{12}(18)^2 = \frac{324}{12} \quad \checkmark$$

Solution to Exercise #64

$$\text{(A) } y = \frac{5x + 60}{x^2 + 2x - 24} = \frac{5x + 60}{(x+6)(x-4)}$$

$$y = \frac{p}{x+6} + \frac{q}{x-4} = \frac{p(x-4) + q(x+6)}{(x+6)(x-4)}$$

$$p(x-4) + q(x+6) = 5x + 60$$

$$px - 4p + qx + 6q = 5x + 60$$

$$p + q = 5$$

$$-4p + 6q = 60$$

$$q = 5 - p$$

$$-4p + 6(5 - p) = 60$$

$$-4p + 30 - 6p = 60$$

$$-10p = 30$$

$$p = -\frac{30}{10} = -3$$

$$q = 5 - p = 5 - (-3) = 5 + 3 = 8$$

$$y = \frac{p}{x+6} + \frac{q}{x-4} = \frac{-3}{x+6} + \frac{8}{x-4}$$

$$y = \frac{a}{bx+c} + \frac{d}{ex+f}$$

$$a = \boxed{-3} \quad, \quad b = \boxed{1} \quad, \quad c = \boxed{6}$$

$$d = \boxed{8} \quad, \quad e = \boxed{1} \quad, \quad f = \boxed{-4}$$

One way to solve this problem is to note that $x^2 + 2x - 24$ can be factored as $(x+6)(x-4)$. Let

$$y = \frac{p}{x+6} + \frac{q}{x-4}$$

where p and q still need to be determined. The way to add fractions is to find a common denominator: $\frac{p}{x+6} + \frac{q}{x-4} = \frac{p(x-4)+q(x+6)}{(x+6)(x-4)} = \frac{px-4p+qx+6q}{(x+6)(x-4)}$.

Comparison with the given equation shows that $p + q = 5$ and $-4p + 6q = 60$. Solve this system to find that $p = -3$ and $q = 8$. Now compare $\frac{-3}{x+6} + \frac{8}{x-4}$ with $\frac{a}{bx+c} + \frac{d}{ex+f}$ to determine a thru f.

An alternative way to solve this problem is to begin with $y = \frac{a}{bx+c} + \frac{d}{ex+f}$, make a common denominator,

(B) If $x = -6$ or if $x = 4$, this would cause division by zero.	and compare with the given equation.

Note: This technique of algebra is called partial fractions. It is useful, for example, in calculus (specifically, one of the techniques of integration), but knowledge of calculus is NOT needed to solve this algebra problem.

Solution to Exercise #65	
(A) $5x = 20 - 4y$ $x = \dfrac{20 - 4y}{5}$ $5\left(\dfrac{20 - 4y}{5}\right)^2 + 4y^2 = 2180$ $5\left(\dfrac{400 - 160y + 16y^2}{25}\right) + 4y^2 = 2180$ $\left(\dfrac{400 - 160y + 16y^2}{5}\right) + 4y^2 = 2180$ $400 - 160y + 16y^2 + 20y^2 = 10{,}900$ $36y^2 - 160y - 10{,}500 = 0$ $9y^2 - 40y - 2625 = 0$ $y = \dfrac{-(-40) \pm \sqrt{(-40)^2 - 4(9)(-2625)}}{2(9)}$ $y = \dfrac{40 \pm \sqrt{1600 + 94{,}500}}{18} = \dfrac{40 \pm \sqrt{96{,}100}}{18}$ $y = \dfrac{40 \pm 310}{18}$ $y = \dfrac{40 + 310}{18}$ or $y = \dfrac{40 - 310}{18}$ $y = \dfrac{350}{18}$ or $y = \dfrac{-270}{18}$ $y = \boxed{\dfrac{175}{9}}$ or $y = \boxed{-15}$	Isolate x in one equation, and plug it into the other equation. Recall that $\left(\dfrac{a}{b}\right)^2 = \dfrac{a^2}{b^2}$ and that $(c - d)^2 = c^2 - 2cd + d^2$. Let $d = 4y$. Then $d^2 = (4y)^2 = 16y^2$. Note that $\dfrac{5}{25} = \dfrac{1}{5}$. Multiply both sides by 5. This calculation is a little simpler to carry out by hand for the student who is proficient in factoring: 25 and 4 may be factored out before doing the arithmetic.

$$x = \frac{20 - 4y}{5}$$

$$x = \frac{20 - 4\left(\frac{175}{9}\right)}{5} \quad \text{or} \quad x = \frac{20 - 4(-15)}{5}$$

$$x = \frac{\frac{180}{9} - \frac{700}{9}}{5} \quad \text{or} \quad x = \frac{20 + 60}{5}$$

$$x = -\frac{520}{45} \quad \text{or} \quad x = \frac{80}{5}$$

$$x = \boxed{-\frac{104}{9}} \quad \text{or} \quad x = \boxed{16}$$

(B) $5x + 4y = 20$

$$5\left(-\frac{104}{9}\right) + 4\left(\frac{175}{9}\right) = -\frac{520}{9} + \frac{700}{9} = \frac{180}{9}$$

$$= 20 \quad \checkmark$$

$$5(16) + 4(-15) = 80 - 60 = 20 \quad \checkmark$$

$$5x^2 + 4y^2 = 2180$$

$$5\left(-\frac{104}{9}\right)^2 + 4\left(\frac{175}{9}\right)^2 = 5\left(\frac{10{,}816}{81}\right) + 4\left(\frac{30{,}625}{81}\right)$$

$$\frac{54{,}080}{81} + \frac{122{,}500}{81} = \frac{176{,}580}{81} = 2180 \quad \checkmark$$

$$5(16)^2 + 4(-15)^2 = 5(256) + 4(225)$$

$$= 1280 + 900 = 2180 \quad \checkmark$$

Solution to Exercise #66	
$$pw = qx$$ $$\frac{w}{x} = \frac{q}{p}$$	Isolate $\frac{w}{x}$ in this equation.
$$ry = sz$$ $$\frac{r}{s} = \frac{z}{y}$$	Isolate $\frac{r}{s}$ in this equation.
$$ry^k = qx^k$$ $$sz^k = pw^k$$	Divide these equations.

$$\frac{ry^k}{sz^k} = \frac{qx^k}{pw^k}$$

$$\frac{zy^k}{yz^k} = \frac{wx^k}{xw^k}$$

$$\frac{y^{k-1}}{z^{k-1}} = \frac{x^{k-1}}{w^{k-1}}$$

$$\left(\frac{y}{z}\right)^{k-1} = \left(\frac{x}{w}\right)^{k-1}$$

$$\boxed{\frac{y}{z} = \frac{x}{w}}$$

Recall $\frac{r}{s} = \frac{z}{y}$ and $\frac{w}{x} = \frac{q}{p}$ from earlier.

$$\frac{y^k}{y} = y^{k-1}, \frac{z}{z^k} = \frac{1}{z^{k-1}}, \text{etc.}$$

Raise both sides of the equation to the power of $\frac{1}{k-1}$ since $(k-1)\left(\frac{1}{k-1}\right) = 1$. (Since $k \neq 1$, this will not involve division by zero.)

Note: The algebra in this problem is similar to calculations for heat engines in thermodynamics. For the Carnot cycle, $P_aV_a = P_bV_b$, $P_bV_b^\gamma = P_cV_c^\gamma$, $P_cV_c = P_dV_d$, and $P_dV_d^\gamma = P_aV_a^\gamma$, which leads to $\frac{V_c}{V_d} = \frac{V_b}{V_a}$.

Solution to Exercise #67

(A) $\frac{1}{w} > \frac{20}{72} - \frac{21}{72} + \frac{2}{72}$

$\frac{1}{x} > \frac{18}{84} - \frac{32}{84} - \frac{7}{84}$

$\frac{1}{y} < \frac{32}{60} - \frac{9}{60} + \frac{7}{60}$

$\frac{1}{z} < \frac{27}{48} - \frac{10}{48} - \frac{23}{48}$

$\frac{1}{w} > \frac{1}{72}$, $\frac{1}{x} > -\frac{21}{84}$

$\frac{1}{y} < \frac{30}{60}$, $\frac{1}{z} < -\frac{6}{48}$

$\frac{1}{w} > \frac{1}{72}$, $\frac{1}{x} > -\frac{1}{4}$

$\frac{1}{y} < \frac{1}{2}$, $\frac{1}{z} < -\frac{1}{8}$

$0 < w < 72$, $x < -4$ or $x > 0$

$y > 2$ or $y < 0$, $0 > z > -8$

(C) $w < 72$

$w = 71$ is < 72

$\frac{1}{w} > \frac{5}{18} - \frac{7}{24} + \frac{1}{36}$

$\frac{1}{71} \approx 0.0141$

$0.0141 > 0.0139$ ✓

$w = 73$ is NOT < 72

$\frac{1}{71} \approx 0.0137$

0.0137 is NOT > 0.0139 ✓

$x < -4$

$x = -4.1$ is < -4

$\frac{1}{x} > \frac{3}{14} - \frac{8}{21} - \frac{1}{12}$

$\frac{1}{-4.1} \approx -0.244$

(B) In the final step, the direction of the inequality reverses regardless of the sign of the variable. Another way to see this is to first cross multiply. As a simple example, $\frac{1}{w} > \frac{1}{72}$ becomes $72 > w$, which is consistent with $w < 72$. As a more involved example, consider $\frac{1}{x} > -\frac{1}{4}$. Any positive value of x will obviously satisfy this inequality, but there are also some negative values of x that will satisfy this inequality. When x is negative, $\frac{1}{x} > -\frac{1}{4}$ becomes $4 < -x$ when cross multiplying. Why? The direction of the inequality changes when multiplying by the negative value of x. Dividing both sides by negative one reverses the direction of the inequality again to get $-4 > x$, which is consistent with $x < -4$. Note that $\frac{1}{x} > -\frac{1}{4}$ and $\frac{1}{y} < \frac{1}{2}$ have two solutions, since any positive value of x and any negative value of y clearly satisfies these inequalities, in addition to the solutions $x < -4$ and $y > 2$.

$$-0.244 > -0.25 \quad \checkmark$$
$$x = -3.9 \text{ is NOT} < -4$$
$$\frac{1}{-3.9} \approx -0.256$$
$$-0.256 \text{ is NOT} > -0.25 \quad \checkmark$$
$$x > 0$$
$$x = 0.1 \text{ is} > 0$$
$$\frac{1}{x} > \frac{3}{14} - \frac{8}{21} - \frac{1}{12}$$
$$\frac{1}{0.1} = 10$$
$$10 > -0.25 \quad \checkmark$$
$$x = -0.1 \text{ is NOT} > 0$$
$$\frac{1}{-0.1} = -10$$
$$-10 \text{ is NOT} > 0 \quad \checkmark$$
$$y > 2$$
$$y = 2.1 \text{ is} > 2$$
$$\frac{1}{y} < \frac{8}{15} - \frac{3}{20} + \frac{7}{60}$$
$$\frac{1}{2.1} \approx 0.476$$
$$0.476 < 0.5 \quad \checkmark$$
$$y = 1.9 \text{ is NOT} > 2$$
$$\frac{1}{1.9} \approx 0.526$$
$$0.526 \text{ is NOT} < 0.5 \quad \checkmark$$
$$y < 0$$
$$y = -0.1 \text{ is} < 0$$
$$\frac{1}{y} < \frac{8}{15} - \frac{3}{20} + \frac{7}{60}$$
$$\frac{1}{-0.1} = -10$$

	$-10 < 0$ ✔
	$y = 0.1$ is NOT < 0
	$\dfrac{1}{0.1} = 10$
	10 is NOT < 0 ✔
	$z > -8$
	$z = -7.9$ is > -8
	$\dfrac{1}{z} < \dfrac{9}{16} - \dfrac{5}{24} - \dfrac{23}{48}$
	$\dfrac{1}{-7.9} \approx -0.127$
	$-0.127 < -0.125$ ✔
	$z = -8.1$ is NOT < -8
	$\dfrac{1}{-8.1} \approx -0.123$
	-0.123 is NOT < -0.125 ✔

Solution to Exercise #68
(A) $t^2 = w^2 + y^2$
$t^2 - y^2 = w^2$
$u^2 = w^2 + z^2 = t^2 - y^2 + z^2$
$x + y = z$
$u^2 = t^2 - y^2 + (x + y)^2$
$u^2 = t^2 - y^2 + x^2 + 2xy + y^2$
$u^2 = t^2 + x^2 + 2xy$
$\boxed{u^2 > t^2 + x^2}$

(B) The problem states that $x > 0$ and $y > 0$. Thus, $2xy > 0$, which ensures that u^2 is greater than $t^2 + x^2$.

Note: A similar inequality arises in geometry in a proof that $c^2 > a^2 + b^2$ if c is the side opposite to the obtuse angle of an obtuse triangle.

Solution to Exercise #69

(A) $\dfrac{x}{y}(2 - z^2) + 2\left(z^2 - \dfrac{x}{y}\right) + z^2\left(\dfrac{x}{y} - 2\right)$

$$= \dfrac{2x}{y} - \dfrac{xz^2}{y} + 2z^2 - \dfrac{2x}{y} + \dfrac{xz^2}{y} - 2z^2 = \boxed{0}$$

(B) $x = 8, y = 2, z = 3$

$$\dfrac{x}{y}(2 - z^2) + 2\left(z^2 - \dfrac{x}{y}\right) + z^2\left(\dfrac{x}{y} - 2\right)$$

$$= \dfrac{8}{2}(2 - 3^2) + 2\left(3^2 - \dfrac{8}{2}\right) + 3^2\left(\dfrac{8}{2} - 2\right)$$

$$= 4(2 - 9) + 2(9 - 4) + 9(4 - 2)$$

$$= 4(-7) + 2(5) + 9(2) = -28 + 10 + 18 = 0 \quad \checkmark$$

(C) Yes. The terms will still cancel out.

Note: This problem is modeled after the following algebraic identity:

$$a(b - c) + b(c - a) + c(a - b) = 0$$

In Part A, $a = \dfrac{x}{y}$, $b = 2$, and $c = z^2$.

Solution to Exercise #70

(A) $(x - 5)^3 = (x - 10)^2(x + 25)$

$$(x - 5)(x^2 - 10x + 25) = (x^2 - 20x + 100)(x + 25)$$

$$x^3 - 10x^2 + 25x - 5x^2 + 50x - 125 = x^3 - 20x^2 + 100x + 25x^2 - 500x + 2500$$

$$x^3 - 15x^2 + 75x - 125 = x^3 + 5x^2 - 400x + 2500$$

$$0 = 20x^2 - 475x + 2625$$

$$0 = 4x^2 - 95x + 525$$

$$x = \dfrac{-(-95) \pm \sqrt{(-95)^2 - 4(4)(525)}}{2(4)}$$

$$x = \dfrac{95 \pm \sqrt{9025 - 8400}}{8} = \dfrac{95 \pm \sqrt{625}}{8} = \dfrac{95 \pm 25}{8}$$

$$x = \dfrac{95 + 25}{8} \quad \text{or} \quad x = \dfrac{95 - 25}{8}$$

$$x = \dfrac{120}{8} \quad \text{or} \quad x = \dfrac{70}{8}$$

$$x = \boxed{15} \quad \text{or} \quad x = \boxed{\frac{35}{4}} = \boxed{8.75}$$

(B) $(x - 5)^3 = (x - 10)^2(x + 25)$

$(15 - 5)^3 = (15 - 10)^2(15 + 25)$

$10^3 = (5)^2(40)$

$1000 = (25)(40) \quad \checkmark$

$$\left(\frac{35}{4} - 5\right)^3 = \left(\frac{35}{4} - 10\right)^2 \left(\frac{35}{4} + 25\right)$$

$$\left(\frac{35}{4} - \frac{20}{4}\right)^3 = \left(\frac{35}{4} - \frac{40}{4}\right)^2 \left(\frac{35}{4} + \frac{100}{4}\right)$$

$$\left(\frac{15}{4}\right)^3 = \left(-\frac{5}{4}\right)^2 \left(\frac{135}{4}\right)$$

$15^3 = (5)^2(135)$

$3375 = (25)(135) \quad \checkmark$

Solution to Exercise #71

(A) $2(x - 3)^2 - 5 = 4x - \dfrac{13}{2}$

$2(x^2 - 6x + 9) - 5 = 4x - \dfrac{13}{2}$

$2x^2 - 12x + 18 - 5 = 4x - \dfrac{13}{2}$

$2x^2 - 16x + 13 + \dfrac{13}{2} = 0$

$4x^2 - 32x + 26 + 13 = 0$

$4x^2 - 32x + 39 = 0$

$(2x - 3)(2x - 13) = 0$

$2x - 3 = 0 \quad \text{or} \quad 2x - 13 = 0$

$2x = 3 \quad \text{or} \quad 2x = 13$

This problem is continued on the next page.

(B) $y = 2(x - 3)^2 - 5$

$$2\left(\frac{3}{2} - 3\right)^2 - 5 = 2\left(\frac{3}{2} - \frac{6}{2}\right)^2 - 5$$

$$= 2\left(-\frac{3}{2}\right)^2 - 5 = 2\left(\frac{9}{4}\right) - 5 = \frac{9}{2} - \frac{10}{2} = -\frac{1}{2} \quad \checkmark$$

$$2\left(\frac{13}{2} - 3\right)^2 - 5 = 2\left(\frac{13}{2} - \frac{6}{2}\right)^2 - 5$$

$$= 2\left(\frac{7}{2}\right)^2 - 5 = 2\left(\frac{49}{4}\right) - 5 = \frac{49}{2} - \frac{10}{2} = \frac{39}{2} \quad \checkmark$$

$$y = 4x - \frac{13}{2}$$

$$4\left(\frac{3}{2}\right) - \frac{13}{2} = \frac{12}{2} - \frac{13}{2} = -\frac{1}{2} \quad \checkmark$$

$$4\left(\frac{13}{2}\right) - \frac{13}{2} = \frac{52}{2} - \frac{13}{2} = \frac{39}{2} \quad \checkmark$$

$$x = \frac{3}{2} \quad \text{or} \quad x = \frac{13}{2}$$

$$y = 4x - \frac{13}{2}$$

$$y = 4\left(\frac{3}{2}\right) - \frac{13}{2} \quad \text{or} \quad y = 4\left(\frac{13}{2}\right) - \frac{13}{2}$$

$$y = \frac{12}{2} - \frac{13}{2} \quad \text{or} \quad y = \frac{52}{2} - \frac{13}{2}$$

$$y = -\frac{1}{2} \quad \text{or} \quad y = \frac{39}{2}$$

$$\left(\frac{3}{2}, -\frac{1}{2}\right) = \boxed{(1.5, -0.5)}$$

$$\text{or} \quad \boxed{\left(\frac{13}{2}, \frac{39}{2}\right)} = \boxed{(6.5, 19.5)}$$

(C) The line has a slope of 4 and a y-intercept of -6.5. The parabola has a minimum of $y = -5$ at $x = 3$ and a y-intercept of 13. (To find the y-intercept of the parabola, set $x = 0$.) The line and parabola intersect at $\left(\frac{3}{2}, -\frac{1}{2}\right)$ and $\left(\frac{13}{2}, \frac{39}{2}\right)$; in decimal form, these coordinates are $(1.5, -0.5)$ and $(6.5, 19.5)$.

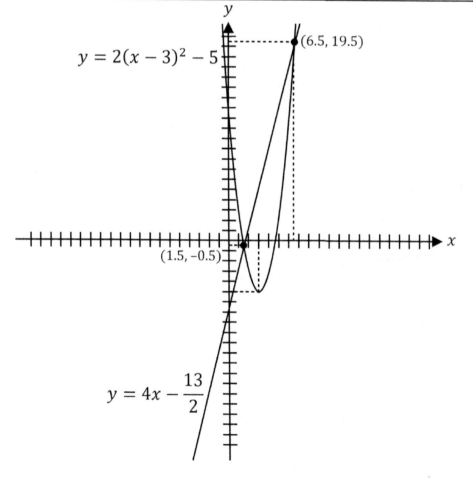

Solution to Exercise #72

(A) $y = \dfrac{1}{\sqrt{1 - \dfrac{u^2}{c^2}}}$

Square both sides of the equation.

$y^2 = \dfrac{1}{1 - \dfrac{u^2}{c^2}} = \dfrac{1}{\dfrac{c^2}{c^2} - \dfrac{u^2}{c^2}}$

Make a common denominator.

$y^2 = \dfrac{1}{\dfrac{c^2 - u^2}{c^2}} = \dfrac{c^2}{c^2 - u^2}$

To divide by a fraction, multiply by its reciprocal: $\dfrac{1}{a/b} = 1 \div \dfrac{a}{b} = 1 \times \dfrac{b}{a} = \dfrac{b}{a}$.

$y^2(c^2 - u^2) = c^2$

Cross multiply.

$w = kyu$

Divide both sides by ky.

$\dfrac{w}{ky} = u$

Plug $u = \dfrac{w}{ky}$ into $y^2(c^2 - u^2) = c^2$.

$y^2 \left(c^2 - \dfrac{w^2}{k^2 y^2} \right) = c^2$

Distribute: $p(q - r) = pq - pr$.

$y^2 c^2 - \dfrac{w^2}{k^2} = c^2$

Multiply both sides by k^2.

$y^2 k^2 c^2 - w^2 = k^2 c^2$

Add w^2 to both sides.

$y^2 k^2 c^2 = w^2 + k^2 c^2$

Multiply both sides by c^2. Note that $c^2 c^2 = c^4$.

$\boxed{y^2 k^2 c^4 = w^2 c^2 + k^2 c^4}$

(B) $\dfrac{y^2}{y^2 - 1} = \dfrac{c^2}{u^2}$

Plug $y^2 = \dfrac{c^2}{c^2 - u^2}$ (from the third line of the solution to Part A) into this equation.

$\dfrac{\dfrac{c^2}{c^2 - u^2}}{\dfrac{c^2}{c^2 - u^2} - 1} = \dfrac{c^2}{u^2}$

Make a common denominator.

$\dfrac{\dfrac{c^2}{c^2 - u^2}}{\dfrac{c^2}{c^2 - u^2} - \dfrac{c^2 - u^2}{c^2 - u^2}} = \dfrac{c^2}{u^2}$

$\dfrac{\dfrac{c^2}{c^2 - u^2}}{\dfrac{c^2 - (c^2 - u^2)}{c^2 - u^2}} = \dfrac{c^2}{u^2}$

Distribute the minus sign:
$-(c^2 - u^2) = -c^2 - (-u^2) = -c^2 + u^2$

$$\frac{\dfrac{c^2}{c^2 - u^2}}{\dfrac{c^2 - c^2 + u^2}{c^2 - u^2}} = \frac{c^2}{u^2}$$

$$\frac{\dfrac{c^2}{c^2 - u^2}}{\dfrac{u^2}{c^2 - u^2}} = \frac{c^2}{u^2} \quad \checkmark$$

To divide by a fraction, multiply by its reciprocal:

$$\frac{c^2}{c^2 - u^2} \div \frac{u^2}{c^2 - u^2} = \frac{c^2}{c^2 - u^2} \times \frac{c^2 - u^2}{u^2}$$

$$= \frac{c^2}{u^2}$$

$$\text{(C)} \ \sqrt{1 - \frac{u^2}{c^2}} > 0$$

This is required because $y > 0$.
Square both sides of the inequality.

$$1 - \frac{u^2}{c^2} > 0$$

Add $\frac{u^2}{c^2}$ to both sides.

$$1 > \frac{u^2}{c^2}$$

Multiply both sides by c^2.

$$c^2 > u^2$$

$$c > u \quad \text{and} \quad u > 0$$

$$\boxed{0 < u < c}$$

Square root both sides. Since $u > 0$, only the positive root applies. Combine $c > u$ with $u > 0$ (which was given) to get $c > u > 0$, which is equivalent to $0 < u < c$. (Note that $u \neq c$ because that would make y infinite.)

$$1 - \frac{u^2}{c^2} < 1$$

$$\sqrt{1 - \frac{u^2}{c^2}} < 1$$

Since $u < c$, $\frac{u}{c} < 1$, such that $1 - \frac{u^2}{c^2} < 1$.

$$y = \frac{1}{\sqrt{1 - \frac{u^2}{c^2}}}$$

$$\boxed{y > 1}$$

Square root both sides. Since $\sqrt{1 - \frac{u^2}{c^2}}$ is < 1, $y = \frac{1}{\sqrt{1 - \frac{u^2}{c^2}}}$ must be > 1.

Note: This problem is similar to formulas from Einstein's special theory of relativity: $\gamma = \frac{1}{\sqrt{1 - \frac{v^2}{c^2}}}$, $p = \gamma m_o v$, $E = \gamma m_o c^2$, and $E^2 = p^2 c^2 + m_0^2 c^4$. The relativistic mass is related to the rest mass by $m = \gamma m_0$, such that $E = mc^2$. When $v = 0$, $\gamma = 1$, so that an object at rest has rest energy according to $E_0 = m_0 c^2$.

Solution to Exercise #73	
$$(A) \sqrt{1 + \sqrt{4 + \sqrt{21 + \sqrt{x + \sqrt{3x^2 - 55x + 267}}}}} = 2$$	Square both sides of the equation: $\sqrt{z} = 2$ becomes $z = 2^2 = 4$.
$$1 + \sqrt{4 + \sqrt{21 + \sqrt{x + \sqrt{3x^2 - 55x + 267}}}} = 2^2 = 4$$	Subtract 1 from both sides.
$$\sqrt{4 + \sqrt{21 + \sqrt{x + \sqrt{3x^2 - 55x + 267}}}} = 4 - 1 = 3$$	Square both sides.
$$4 + \sqrt{21 + \sqrt{x + \sqrt{3x^2 - 55x + 267}}} = 3^2 = 9$$	Subtract 4 from both sides.
$$\sqrt{21 + \sqrt{x + \sqrt{3x^2 - 55x + 267}}} = 9 - 4 = 5$$	Square both sides.
$$21 + \sqrt{x + \sqrt{3x^2 - 55x + 267}} = 5^2 = 25$$	Subtract 21 from both sides.
$$\sqrt{x + \sqrt{3x^2 - 55x + 267}} = 25 - 21 = 4$$	Square both sides.
$$x + \sqrt{3x^2 - 55x + 267} = 4^2 = 16$$	Subtract x from both sides.
$$\sqrt{3x^2 - 55x + 267} = 16 - x$$	Square both sides.
$$3x^2 - 55x + 267 = (16 - x)^2 = 256 - 32x + x^2$$	Combine like terms.
$$2x^2 - 23x + 11 = 0$$	Factor the quadratic, or use
$$(2x - 1)(x - 11) = 0$$	the quadratic formula.
$$2x - 1 = 0 \quad \text{or} \quad x - 11 = 0$$	
$$2x = 1 \quad \text{or} \quad x = 11$$	There are two possible
$$x = \boxed{\frac{1}{2}} = \boxed{0.5} \quad \text{or} \quad x = \boxed{11}$$	answers.

(B) $\sqrt{1+\sqrt{4+\sqrt{21+\sqrt{\frac{1}{2}+\sqrt{3\left(\frac{1}{2}\right)^2-55\left(\frac{1}{2}\right)+267}}}}}$ Plug in $x=\frac{1}{2}$.

$=\sqrt{1+\sqrt{4+\sqrt{21+\sqrt{\frac{1}{2}+\sqrt{\frac{3}{4}-\frac{55}{2}+267}}}}}$ Make a common denominator.

$=\sqrt{1+\sqrt{4+\sqrt{21+\sqrt{\frac{1}{2}+\sqrt{\frac{3}{4}-\frac{110}{4}+\frac{1068}{4}}}}}}$ Note that $31^2=961$.

$=\sqrt{1+\sqrt{4+\sqrt{21+\sqrt{\frac{1}{2}+\sqrt{\frac{961}{4}}}}}}$

$=\sqrt{1+\sqrt{4+\sqrt{21+\sqrt{\frac{1}{2}+\frac{31}{2}}}}}$

$=\sqrt{1+\sqrt{4+\sqrt{21+\sqrt{16}}}}$

$=\sqrt{1+\sqrt{4+\sqrt{21+4}}}=\sqrt{1+\sqrt{4+\sqrt{25}}}$

$=\sqrt{1+\sqrt{4+5}}=\sqrt{1+\sqrt{9}}=\sqrt{1+3}=\sqrt{4}=2$ ✓

$$\sqrt{1 + \sqrt{4 + \sqrt{21 + \sqrt{11 + \sqrt{3(11)^2 - 55(11) + 267}}}}}$$

$$= \sqrt{1 + \sqrt{4 + \sqrt{21 + \sqrt{11 + \sqrt{3(121) - 605 + 267}}}}}$$

$$= \sqrt{1 + \sqrt{4 + \sqrt{21 + \sqrt{11 + \sqrt{363 - 338}}}}}$$

$$= \sqrt{1 + \sqrt{4 + \sqrt{21 + \sqrt{11 + \sqrt{25}}}}}$$

$$= \sqrt{1 + \sqrt{4 + \sqrt{21 + \sqrt{11 + 5}}}}$$

$$= \sqrt{1 + \sqrt{4 + \sqrt{21 + \sqrt{16}}}} = \sqrt{1 + \sqrt{4 + \sqrt{21 + 4}}}$$

$$= \sqrt{1 + \sqrt{4 + \sqrt{25}}} = \sqrt{1 + \sqrt{4 + 5}}$$

$$= \sqrt{1 + \sqrt{9}} = \sqrt{1 + 3} = \sqrt{4} = 2 \quad \checkmark$$

Solution to Exercise #74

(A) $12\left(\dfrac{2x}{3}-\dfrac{3y}{4}\right)=12\left(\dfrac{1}{6}\right)$	Multiply both sides of the first equation by 12.
$4\left(\dfrac{5x}{7}+\dfrac{9y}{4}\right)=4\left(\dfrac{17}{4}\right)$	Multiply both sides of the second equation by 4.
$8x-9y=2$	Add the two equations together: $9y$ cancels out.
$\dfrac{20x}{7}+9y=17$	
$8x+\dfrac{20x}{7}=19$	Multiply both sides of the equation by 7. (This way it is not necessary to make a common denominator.)
$56x+20x=133$	
$76x=133$	
$x=\dfrac{133}{76}=\boxed{\dfrac{7}{4}}$	Divide 133 and 76 each by 19 to see that $\dfrac{133}{76}$ reduces to $\dfrac{7}{4}$.
$\dfrac{5}{7}\left(\dfrac{7}{4}\right)+\dfrac{9y}{4}=\dfrac{17}{4}$	
$\dfrac{5}{4}+\dfrac{9y}{4}=\dfrac{17}{4}$	Multiply both sides of the equation by 4.
$5+9y=17$	
$9y=12$	
$y=\dfrac{12}{9}=\boxed{\dfrac{4}{3}}$	

(B) $\dfrac{2}{3}\left(\dfrac{7}{4}\right)-\dfrac{3}{4}\left(\dfrac{4}{3}\right)=\dfrac{7}{6}-1=\dfrac{7}{6}-\dfrac{6}{6}=\dfrac{1}{6}$ ✓

$\dfrac{5}{7}\left(\dfrac{7}{4}\right)+\dfrac{9}{4}\left(\dfrac{4}{3}\right)=\dfrac{5}{4}+\dfrac{12}{4}=\dfrac{17}{4}$ ✓

Solution to Exercise #75

(A) $p^2x^2=c^2k^2$	Square both sides of this equation.
$p^2y^2=\dfrac{k^2u^4}{w^2}$	Also square both sides of this equation. Add the two squared equations together.
$p^2x^2+p^2y^2=c^2k^2+\dfrac{k^2u^4}{w^2}$	Factor out p^2. The reason for squaring equations and adding them together is
$p^2(x^2+y^2)=c^2k^2+\dfrac{k^2u^4}{w^2}$	to take advantage of $x^2+y^2=1$.

(B) $6z - 4y = 3$

$$6\left(\frac{128}{243}\right) - 4\left(\frac{13}{324}\right) = \frac{256}{81} - \frac{13}{81} = \frac{243}{81}$$

$$= 3 \quad \checkmark$$

$$6\left(\frac{2}{3}\right) - 4\left(\frac{1}{4}\right) = 4 - 1 = 3 \quad \checkmark$$

$$3y = x - \frac{1}{4}$$

$$3\left(\frac{13}{324}\right) = \frac{13}{108} = \frac{10}{27} - \frac{1}{4} = \frac{40}{108} - \frac{27}{108}$$

$$= \frac{13}{108} \quad \checkmark$$

$$3\left(\frac{1}{4}\right) = \frac{3}{4} = 1 - \frac{1}{4} = \frac{3}{4} \quad \checkmark$$

$$\frac{z}{6} = \left(x - \frac{2}{3}\right)^2$$

$$\frac{1}{6}\left(\frac{128}{243}\right) = \frac{64}{729}$$

$$\left(\frac{10}{27} - \frac{2}{3}\right)^2 = \left(\frac{10}{27} - \frac{18}{27}\right)^2 = \left(-\frac{8}{27}\right)^2$$

$$= \frac{64}{729} \quad \checkmark$$

$$\frac{1}{6}\left(\frac{2}{3}\right) = \frac{1}{9} = \left(1 - \frac{2}{3}\right)^2 = \left(\frac{1}{3}\right)^2 = \frac{1}{9} \quad \checkmark$$

Solution to Exercise #81

(A) $\dfrac{3x+2}{2x-1}\left(\dfrac{6x+8}{6x+8}\right)-\dfrac{4x+5}{6x+8}\left(\dfrac{2x-1}{2x-1}\right)+\dfrac{2}{x}$

$=\dfrac{18x^2+24x+12x+16}{12x^2+16x-6x-8}$

$\quad-\dfrac{8x^2-4x+10x-5}{12x^2-6x+16x-8}+\dfrac{2}{x}$

$=\dfrac{18x^2+36x+16}{12x^2+10x-8}-\dfrac{8x^2+6x-5}{12x^2+10x-8}+\dfrac{2}{x}$

$=\dfrac{18x^2+36x+16-(8x^2+6x-5)}{12x^2+10x-8}+\dfrac{2}{x}$

$=\dfrac{18x^2+36x+16-8x^2-6x+5}{12x^2+10x-8}+\dfrac{2}{x}$

$=\dfrac{10x^2+30x+21}{12x^2+10x-8}+\dfrac{2}{x}$

$=\dfrac{10x^2+30x+21}{12x^2+10x-8}\left(\dfrac{x}{x}\right)+\dfrac{2}{x}\left(\dfrac{12x^2+10x-8}{12x^2+10x-8}\right)$

$=\dfrac{10x^3+30x^2+21x}{12x^3+10x^2-8x}+\dfrac{24x^2+20x-16}{12x^3+10x^2-8x}$

$=\boxed{\dfrac{10x^3+54x^2+41x-16}{12x^3+10x^2-8x}}=\dfrac{P}{Q}$

$P=\boxed{10x^3+54x^2+41x-16}$

$Q=\boxed{12x^3+10x^2-8x}$

(B) $\dfrac{10(2)^3+54(2)^2+41(2)-16}{12(2)^3+10(2)^2-8(2)}$

$=\dfrac{10(8)+54(4)+82-16}{12(8)+10(4)-16}$

$=\dfrac{80+216+66}{96+40-16}$

$=\dfrac{362}{120}=\boxed{\dfrac{181}{60}}$

$\dfrac{3(2)+2}{2(2)-1}-\dfrac{4(2)+5}{6(2)+8}+\dfrac{2}{2}$

$=\dfrac{6+2}{4-1}-\dfrac{8+5}{12+8}+1$

$=\dfrac{8}{3}-\dfrac{13}{20}+1=\dfrac{160}{60}-\dfrac{39}{60}+\dfrac{60}{60}$

$=\boxed{\dfrac{181}{60}}\quad\checkmark$

(C) $2x-1=0$,

$\qquad 6x+8=0$, $\quad x=0$

$2x=1$, $\quad 6x=-8$, $\quad x=0$

$x=\boxed{\dfrac{1}{2}}, x=-\dfrac{8}{6}=\boxed{-\dfrac{4}{3}}, x=\boxed{0}$

Solution to Exercise #82

(A) $(x-3)^2+(y-8)^2=25$

$3y=4x+12$

$y=\dfrac{4x}{3}+\dfrac{12}{3}=\dfrac{4x}{3}+4$

$(x-3)^2+\left(\dfrac{4x}{3}+4-8\right)^2=25$

$(x-3)^2+\left(\dfrac{4x}{3}-4\right)^2=25$

(B) $\sqrt{(x-3)^2+(y-8)^2}=5$

$\sqrt{(0-3)^2+(4-8)^2}=\sqrt{9+16}$

$=\sqrt{25}=5\quad\checkmark$

$\sqrt{(6-3)^2+(12-8)^2}=\sqrt{9+16}$

$=\sqrt{25}=5\quad\checkmark$

$3y=4x+12$

$$x^2 - 6x + 9 + \frac{16x^2}{9} - \frac{32x}{3} + 16 = 25$$

$$x^2 - 6x + \frac{16x^2}{9} - \frac{32x}{3} = 0$$

$$9x^2 - 54x + 16x^2 - 96x = 0$$

$$25x^2 - 150x = 0$$

$$25x(x - 6) = 0$$

$$25x = 0 \quad \text{or} \quad x - 6 = 0$$

$$x = \boxed{0} \quad \text{or} \quad x = \boxed{6}$$

$$y = \frac{4x}{3} + 4$$

$$y = \frac{4(0)}{3} + 4 \quad \text{or} \quad y = \frac{4(6)}{3} + 4$$

$$y = 0 + 4 \quad \text{or} \quad y = \frac{24}{3} + 4$$

$$y = \boxed{4} \quad \text{or} \quad y = 8 + 4 = \boxed{12}$$

$$3(4) = 12 = 4(0) + 12$$
$$= 0 + 12 = 12 \quad \checkmark$$
$$3(12) = 36 = 4(6) + 12$$
$$= 24 + 12 = 36 \quad \checkmark$$

(C) $3y = 4x + 12$, which is equivalent to $y = \frac{4x}{3} + 4$, represents a straight line with a slope of $\frac{4}{3}$ and a y-intercept of 4, while $\sqrt{(x - 3)^2 + (y - 8)^2} = 5$ represents a circle with a radius of 5 centered about the point $(3, 8)$. The solution to this problem shows that the line and circle intersect at the points $(0, 4)$ and $(6, 12)$.

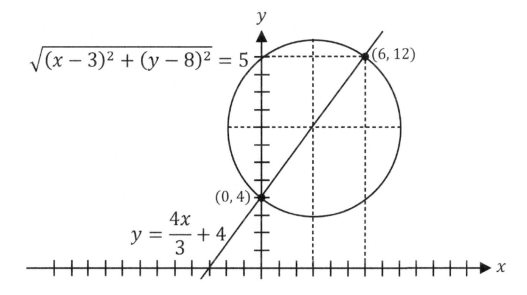

$\sqrt{(x - 3)^2 + (y - 8)^2} = 5$

$(6, 12)$

$(0, 4)$

$y = \frac{4x}{3} + 4$

Solution to Exercise #83	
(A) $ax + by = 0$	Subtract by from both sides.
$ax = -by$	
$z = x - y$	Multiply both sides by b.
$bz = bx - by$	Replace $-by$ with ax (since $-by = ax$).
$bz = bx + ax$	Factor out the x.
$bz = x(b + a)$	Divide both sides by $(b + a)$.
$\dfrac{bz}{b + a} = \boxed{\dfrac{bz}{a + b}} = x$	Note that $b + a = a + b$.
$z = x - y$	Multiply both sides by a.
$az = ax - ay$	Replace ax with $-by$ (since $-by = ax$).
$az = -by - ay$	Factor out the $-y$.
$az = -y(b + a)$	Divide both sides by $(b + a)$.
$\dfrac{-az}{b + a} = \boxed{\dfrac{-az}{a + b}} = y$	Note that $b + a = a + b$.
(B) $w = ax^2 + by^2$	
$w = a\left(\dfrac{bz}{a + b}\right)^2 + b\left(\dfrac{-az}{a + b}\right)^2$	Plug $a = \dfrac{bz}{a+b}$ and $y = \dfrac{-az}{a+b}$ into the equation for w.
$w = \dfrac{ab^2z^2}{(a + b)^2} + \dfrac{ba^2z^2}{(a + b)^2} = \dfrac{abz^2(b + a)}{(a + b)^2}$	Note that $abz^2(b + a) = ab^2z^2 + ba^2z^2$.
$w = \dfrac{abz^2}{a + b}$	Note that $\dfrac{(b+a)}{(b+a)^2} = \dfrac{1}{(b+a)}$ just like $\dfrac{p}{p^2} = \dfrac{1}{p}$.
$\dfrac{1}{a} + \dfrac{1}{b} = \dfrac{1}{c}$	Make a common denominator:
$\dfrac{b + a}{ab} = \dfrac{1}{c}$	$\dfrac{1}{a} + \dfrac{1}{b} = \dfrac{1}{a}\dfrac{b}{b} + \dfrac{1}{b}\dfrac{a}{a} = \dfrac{b}{ab} + \dfrac{a}{ba} = \dfrac{b + a}{ab}$
	Take the reciprocal of each side.
$\dfrac{ab}{a + b} = c$	Replace $\dfrac{ab}{a+b}$ with c in $w = \dfrac{abz^2}{a+b}$.
$\boxed{w = cz^2}$	

Solution to Exercise #84

(A) $3\sqrt{xy} = 28$

$$\sqrt{y} = \frac{28}{3\sqrt{x}}$$

$$8\sqrt{x} - 9\sqrt{y} = 4$$

$$8\sqrt{x} - 9\left(\frac{28}{3\sqrt{x}}\right) = 4$$

$$8\sqrt{x} - \frac{84}{\sqrt{x}} = 4$$

$$8x - 84 = 4\sqrt{x}$$

$$4x - 42 = 2\sqrt{x}$$

$$4x - 2\sqrt{x} - 42 = 0$$

$$\sqrt{x} = \frac{-(-2) \pm \sqrt{(-2)^2 - 4(4)(-42)}}{2(4)}$$

$$\sqrt{x} = \frac{2 \pm \sqrt{4 + 672}}{8} = \frac{2 \pm \sqrt{676}}{8} = \frac{2 \pm 26}{8}$$

$$\sqrt{x} = \frac{2 - 26}{8} \quad \text{or} \quad \sqrt{x} = \frac{2 + 26}{8}$$

$$\sqrt{x} = \frac{-24}{8} \quad \text{or} \quad \sqrt{x} = \frac{28}{8}$$

$$\sqrt{x} = -3 \quad \text{or} \quad \sqrt{x} = \frac{7}{2}$$

$$\text{only} \quad \sqrt{x} = \frac{7}{2} > 0$$

$$x = \left(\frac{7}{2}\right)^2 = \boxed{\frac{49}{4}}$$

$$\sqrt{y} = \frac{28}{3\sqrt{x}} = \frac{28}{3\left(\frac{7}{2}\right)} = \frac{28}{\frac{21}{2}}$$

$$\sqrt{y} = 28 \div \frac{21}{2} = 28 \times \frac{2}{21} = \frac{56}{21} = \frac{8}{3}$$

$$y = \left(\frac{8}{3}\right)^2 = \boxed{\frac{64}{9}}$$

Note that $\sqrt{xy} = \sqrt{x}\sqrt{y}$. Divide both sides by $3\sqrt{x}$.

Plug $\sqrt{y} = \frac{28}{3\sqrt{x}}$ into $8\sqrt{x} - 9\sqrt{y} = 4$. Note that $\frac{9(28)}{3} = 3(28) = 84$.

Multiply both sides by $\sqrt{x}$.

This equation is quadratic in $\sqrt{x}$. If it helps, define $u = \sqrt{x}$, solve the quadratic equation for u, and then use $u^2 = x$ to find x.

The problem states that $\sqrt{x} > 0$, so only $\sqrt{x} = \frac{7}{2}$ applies. Square both sides to find x.

(B) $3\sqrt{xy} = 3\sqrt{\left(\dfrac{49}{4}\right)\left(\dfrac{64}{9}\right)} = 3\left(\dfrac{7}{2}\right)\left(\dfrac{8}{3}\right) = 28$ ✔

$$8\sqrt{x} - 9\sqrt{y} = 8\sqrt{\dfrac{49}{4}} - 9\sqrt{\dfrac{64}{9}}$$

$$= 8\left(\dfrac{7}{2}\right) - 9\left(\dfrac{8}{3}\right) = 28 - 24 = 4$$ ✔

Solution to Exercise #85

(A) $xyz\left(\dfrac{1}{x} + \dfrac{1}{y} + \dfrac{1}{z}\right) = \dfrac{xyz}{x} + \dfrac{xyz}{y} + \dfrac{xyz}{z} = yz + xz + xy$

Divide both sides by xyz.

$$\left(\dfrac{1}{x} + \dfrac{1}{y} + \dfrac{1}{z}\right) = \dfrac{yz + xz + xy}{xyz}$$

Multiply both sides by $(x + y + z)$.

$$(x + y + z)\left(\dfrac{1}{x} + \dfrac{1}{y} + \dfrac{1}{z}\right) = (x + y + z)\dfrac{yz + xz + xy}{xyz}$$

$$(x + y + z)\left(\dfrac{1}{x} + \dfrac{1}{y} + \dfrac{1}{z}\right) = \dfrac{xyz + zx^2 + yx^2 + zy^2 + xyz + xy^2 + yz^2 + xz^2 + xyz}{xyz}$$

$$(x + y + z)\left(\dfrac{1}{x} + \dfrac{1}{y} + \dfrac{1}{z}\right) = \dfrac{(xy^2 + yz^2 + zx^2) + (xz^2 + yx^2 + zy^2) + 3xyz}{xyz}$$

$$xy^2 + yz^2 + zx^2 = a$$

$$xz^2 + yx^2 + zy^2 = b$$

$$xyz = c$$

$$\boxed{(x + y + z)\left(\dfrac{1}{x} + \dfrac{1}{y} + \dfrac{1}{z}\right) = \dfrac{a + b + 3c}{xyz}}$$

(B) $2(3)^2 + 3(6)^2 + 6(2)^2 = 2(9) + 3(36) + 6(4) = 18 + 108 + 24 = \boxed{150} = a$

$2(6)^2 + 3(2)^2 + 6(3)^2 = 2(36) + 3(4) + 6(9) = 72 + 12 + 54 = \boxed{138} = b$

$(2)(3)(6) = \boxed{36} = c$

(C) $(2 + 3 + 6)\left(\dfrac{1}{2} + \dfrac{1}{3} + \dfrac{1}{6}\right) = (11)\left(\dfrac{3}{6} + \dfrac{2}{6} + \dfrac{1}{6}\right) = 11\left(\dfrac{6}{6}\right) = 11(1) = \boxed{11}$

$\dfrac{a + b + 3c}{xyz} = \dfrac{150 + 138 + 3(36)}{(2)(3)(6)} = \dfrac{288 + 108}{36} = \dfrac{396}{36} = \boxed{11}$ ✔

Solution to Exercise #86	
(A) $4x^2 + 6xy + \dfrac{9y^2}{4} = 5y^2 + \dfrac{9y^4}{4}$	Add $\dfrac{9y^2}{4}$ to both sides. Why? This completes the square. (Recall Exercise 30.) The idea is that $(t+u)^2 = t^2 + 2tu + u^2$. Let $t = 2x$ and $2tu = 6xy$ so that $u = \dfrac{3xy}{t} = \dfrac{3xy}{2x} = \dfrac{3y}{2}$. Then $u^2 = \dfrac{9y^2}{4}$.
$\left(2x + \dfrac{3y}{2}\right)^2 = \dfrac{20y^2}{4} + \dfrac{9y^2}{4}$	
$\left(2x + \dfrac{3y}{2}\right)^2 = \dfrac{29y^2}{4}$	
$\left(2x + \dfrac{3y}{2}\right)^2 = \left(\dfrac{y\sqrt{29}}{2}\right)^2$	
$2x + \dfrac{3y}{2} = \pm\dfrac{y\sqrt{29}}{2}$	Square root both sides. Include the positive and negative roots.
$4x + 3y = \pm y\sqrt{29}$	Multiply both sides by 2.
$4x = -3y \pm y\sqrt{29}$	Subtract $3y$ from both sides.
$4x = y(-3 \pm \sqrt{29})$	Factor out the y.
$\dfrac{x}{y} = \boxed{\dfrac{-3 \pm \sqrt{29}}{4}}$	Divide both sides by 4 and divide both sides by y.
(B) $x = \dfrac{(-3 \pm \sqrt{29})y}{4}$	
$4x^2 + 6xy = 5y^2$	
$4\left[\dfrac{(-3 \pm \sqrt{29})y}{4}\right]^2 + 6\left[\dfrac{(-3 \pm \sqrt{29})y}{4}\right]y = 5y^2$	Apply the rule $(py)^2 = p^2y^2$.
$4\left(\dfrac{-3 \pm \sqrt{29}}{4}\right)^2 y^2 + \dfrac{6(-3 \pm \sqrt{29})y^2}{4} = 5y^2$	Divide both sides by y^2; it cancels out.
$4\left(\dfrac{-3 \pm \sqrt{29}}{4}\right)^2 + \dfrac{6(-3 \pm \sqrt{29})}{4} = 5$	Apply the rule $\left(\dfrac{q}{4}\right)^2 = \dfrac{q^2}{4^2} = \dfrac{q^2}{16}$.
$\dfrac{4(-3 \pm \sqrt{29})^2}{16} + \dfrac{6(-3 \pm \sqrt{29})}{4} = 5$	Note that $\dfrac{4}{16} = \dfrac{1}{4}$.
$\dfrac{(-3 \pm \sqrt{29})^2}{4} + \dfrac{6(-3 \pm \sqrt{29})}{4} = 5$	Multiply both sides by 4.

$$\left(-3 \pm \sqrt{29}\right)^2 + 6\left(-3 \pm \sqrt{29}\right) = 20$$
$$(-3)(-3) \mp 3\sqrt{29} \mp 3\sqrt{29} + \sqrt{29}\sqrt{29} - 18$$
$$\pm 6\sqrt{29} = 20$$
$$9 \mp 6\sqrt{29} + 29 - 18 \pm 6\sqrt{29} = 20$$
$$38 - 18 = 20 \quad \checkmark$$

Note that $(-c + d)^2 = c^2 - 2cd + d^2$. The notation $\mp$ means to use the opposite sign compared to $\pm$, such that $\mp 6\sqrt{29} \pm 6\sqrt{29} = 0$.

Solution to Exercise #87

(A) $V_{\text{sphere}} = V_{\text{cube}}$

$$\frac{4}{3}\pi R^3 = L^3$$

$$R^3 = \frac{3L^3}{4\pi}$$

$$R = \left(\frac{3L^3}{4\pi}\right)^{\frac{1}{3}} = (L^3)^{\frac{1}{3}}\left(\frac{3}{4\pi}\right)^{\frac{1}{3}}$$

$$\boxed{R = L\left(\frac{3}{4\pi}\right)^{\frac{1}{3}}} = \boxed{L\left(\sqrt[3]{\frac{3}{4\pi}}\right)}$$

(B) $\dfrac{S_{\text{sphere}}}{S_{\text{cube}}} = \dfrac{4\pi R^2}{6L^2} = \dfrac{2\pi R^2}{3L^2}$

$$\frac{S_{\text{sphere}}}{S_{\text{cube}}} = \frac{2\pi}{3L^2} L^2 \left(\frac{3}{4\pi}\right)^{\frac{2}{3}} = \frac{2\pi}{3}\left(\frac{3}{4\pi}\right)^{\frac{2}{3}}$$

$$\frac{S_{\text{sphere}}}{S_{\text{cube}}} = \frac{2\pi}{3}\frac{3^{2/3}}{4^{2/3}\pi^{2/3}} = \frac{2}{4^{2/3}}\frac{3^{2/3}}{3}\frac{\pi}{\pi^{2/3}}$$

Notes: $\dfrac{3^{2/3}}{3} = \dfrac{1}{3^{1-2/3}} = \dfrac{1}{3^{1/3}}$ and $4^n = (2^2)^n = 2^{2n}$.

$$\frac{S_{\text{sphere}}}{S_{\text{cube}}} = \frac{2}{(2^2)^{2/3}}\frac{1}{3^{1/3}}\pi^{1/3} = \frac{2}{2^{4/3}}\frac{\pi^{1/3}}{3^{1/3}}$$

Note: $\dfrac{2}{2^{4/3}} = \dfrac{1}{2^{4/3-1}} = \dfrac{1}{2^{1/3}}$.

$$\frac{S_{\text{sphere}}}{S_{\text{cube}}} = \frac{1}{2^{1/3}}\frac{\pi^{1/3}}{3^{1/3}} = \left(\frac{\pi}{2\cdot3}\right)^{1/3} = \boxed{\left(\frac{\pi}{6}\right)^{1/3}} = \boxed{\sqrt[3]{\frac{\pi}{6}}}$$

(C) Since $\pi \approx 3.14$, the ratio $\frac{\pi}{6}$ is less than 1. Therefore, $\frac{S_{\text{sphere}}}{S_{\text{cube}}} < 1$ and $S_{\text{sphere}} < S_{\text{cube}}$. If a sphere and cube have the same volume, the sphere has less surface area than the cube. (For a given volume, the sphere is the shape that minimizes the surface area.)

Solution to Exercise $^{\#}88$	
(A) $\dfrac{x}{z} = \dfrac{w}{x}$	Cross multiply: $\dfrac{a}{b} = \dfrac{c}{d}$
$x^2 = wz$	becomes $ad = bc$.
$\dfrac{y}{z} = \dfrac{z-w}{y}$	Cross multiply with this
	equation, too.
$y^2 = z(z-w)$	Distribute the z.
$y^2 = z^2 - wz$	Recall $x^2 = wz$ from earlier.
$y^2 = z^2 - x^2$	Add x^2 to both sides.
$\boxed{x^2 + y^2 = z^2}$	

$$(B)\ \frac{x}{z} = \frac{w}{x}$$

$$\frac{x}{25} = \frac{9}{x}$$

$$x^2 = 9(25) = 225$$

$$x = \pm\sqrt{225} = \boxed{\pm 15}$$

$$\frac{y}{z} = \frac{z-w}{y}$$

$$\frac{y}{25} = \frac{25-9}{y}$$

$$\frac{y}{25} = \frac{16}{y}$$

$$y^2 = 25(16) = 400$$

$$y = \pm\sqrt{400} = \boxed{\pm 20}$$

(C) $(\pm 15)^2 + (\pm 20)^2 = 225 + 400 = 625 = 25^2$ ✔

Note: This problem is modeled after a method of proving the Pythagorean theorem which arises from similar triangles.

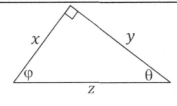

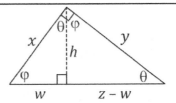

Solution to Exercise #89

(A) $\left(256 + \left(2 + \left(7 + \left(3 + (2x^2 - 27x + 44)^{\frac{1}{3}}\right)^2\right)^{\frac{1}{5}}\right)^4\right)^{\frac{1}{3}} = 8$	Raise both sides to the power of 3.
$256 + \left(2 + \left(7 + \left(3 + (2x^2 - 27x + 44)^{\frac{1}{3}}\right)^2\right)^{\frac{1}{5}}\right)^4 = 8^3 = 512$	Subtract 256 from both sides.
$\left(2 + \left(7 + \left(3 + (2x^2 - 27x + 44)^{\frac{1}{3}}\right)^2\right)^{\frac{1}{5}}\right)^4 = 512 - 256 = 256$	Raise both sides to the power of $\frac{1}{4}$. Only the positive root will lead to a real answer for x.
$2 + \left(7 + \left(3 + (2x^2 - 27x + 44)^{\frac{1}{3}}\right)^2\right)^{\frac{1}{5}} = 256^{1/4} = 4$	
$\left(7 + \left(3 + (2x^2 - 27x + 44)^{\frac{1}{3}}\right)^2\right)^{\frac{1}{5}} = 4 - 2 = 2$	Subtract 2 from both sides. Raise both sides to the power of 5.
$7 + \left(3 + (2x^2 - 27x + 44)^{\frac{1}{3}}\right)^2 = 2^5 = 32$	Subtract 7 from both sides. Square root both sides. Only the positive root will lead to a real answer for x.
$\left(3 + (2x^2 - 27x + 44)^{\frac{1}{3}}\right)^2 = 32 - 7 = 25$	
$3 + (2x^2 - 27x + 44)^{\frac{1}{3}} = \sqrt{25} = 5$	
$(2x^2 - 27x + 44)^{\frac{1}{3}} = 5 - 3 = 2$	
$2x^2 - 27x + 44 = 2^3 = 8$	
$2x^2 - 27x + 36 = 0$	
$(2x - 3)(x - 12) = 0$	
$2x - 3 = 0 \quad \text{or} \quad x - 12 = 0$	
$2x = 3 \quad \text{or} \quad x = 12$	
$x = \boxed{\dfrac{3}{2}} = \boxed{1.5} \quad \text{or} \quad x = \boxed{12}$	
(B) $2x^2 - 27x + 44 = 2\left(\dfrac{3}{2}\right)^2 - 27\left(\dfrac{3}{2}\right) + 44$	First evaluate this expression.

$$= 2\left(\frac{9}{4}\right) - \frac{81}{2} + 44 = \frac{18}{4} - \frac{162}{4} + \frac{176}{4} = \frac{32}{4} = 8$$

$$2x^2 - 27x + 44 = 2(12)^2 - 27(12) + 44 = 2(144) - 324 + 44$$

$$= 288 - 280 = 8$$

Observe that $2x^2 - 27x + 44 = 8$ for each possible value of x. Replace this expression with 8 in the given equation.

$$\left(256 + \left(2 + \left(7 + \left(3 + (2x^2 - 27x + 44)^{\frac{1}{3}}\right)^2\right)^{\frac{1}{5}}\right)^4\right)^{\frac{1}{3}} = 8$$

$$\left(256 + \left(2 + \left(7 + \left(3 + 8^{1/3}\right)^2\right)^{\frac{1}{5}}\right)^4\right)^{\frac{1}{3}}$$

$$= \left(256 + \left(2 + \left(7 + (3 + 2)^2\right)^{\frac{1}{5}}\right)^4\right)^{\frac{1}{3}}$$

$$= \left(256 + \left(2 + \left(7 + 5^2\right)^{\frac{1}{5}}\right)^4\right)^{\frac{1}{3}} = \left(256 + \left(2 + (7 + 25)^{\frac{1}{5}}\right)^4\right)^{\frac{1}{3}}$$

$$= \left(256 + \left(2 + 32^{1/5}\right)^4\right)^{\frac{1}{3}} = \left(256 + (2 + 2)^4\right)^{\frac{1}{3}} = (256 + 4^4)^{\frac{1}{3}}$$

$$= (256 + 256)^{1/3} = 512^{1/3} = \boxed{8}$$

Solution to Exercise #90

$$\text{(A)} \quad \frac{t}{b(y - t)} + \frac{u}{a(u - x)} - \frac{x(t - u)}{a(u - x)^2} = 0$$

$$\frac{t}{b(y - t)} + \frac{u(u - x)}{a(u - x)^2} - \frac{x(t - u)}{a(u - x)^2} = 0$$

$$\frac{t}{b(y - t)} + \frac{u^2 - ux}{a(u - x)^2} - \frac{tx - ux}{a(u - x)^2} = 0$$

$$\frac{t}{b(y - t)} + \frac{u^2 - ux - (tx - ux)}{a(u - x)^2} = 0$$

$$\frac{t}{b(y - t)} + \frac{u^2 - ux - tx + ux}{a(u - x)^2} = 0$$

$$\frac{t}{b(y - t)} + \frac{u^2 - tx}{a(u - x)^2} = 0$$

Make the middle term and the last term have a common denominator.

Distribute the minus sign: $-(tx - ux) = -tx - (-ux) = -tx + ux$.

Subtract $\frac{u^2 - tx}{a(u - x)^2}$ from both sides.

$$\frac{t}{b(y-t)} = \frac{-u^2 + tx}{a(u-x)^2}$$

$$\frac{a(u-x)}{b(y-t)} = \frac{-u^2 + tx}{t(u-x)}$$

$$\frac{t}{b(y-t)} + \frac{u}{a(u-x)} - \frac{y(t-u)}{b(y-t)^2} = 0$$

$$\frac{t(y-t)}{b(y-t)^2} + \frac{u}{a(u-x)} - \frac{y(t-u)}{b(y-t)^2} = 0$$

$$\frac{ty - t^2}{b(y-t)^2} + \frac{u}{a(u-x)} - \frac{ty - uy}{b(y-t)^2} = 0$$

$$\frac{ty - t^2 - (ty - uy)}{b(y-t)^2} + \frac{u}{a(u-x)} = 0$$

$$\frac{ty - t^2 - ty + uy}{b(y-t)^2} + \frac{u}{a(u-x)} = 0$$

$$\frac{-t^2 + uy}{b(y-t)^2} + \frac{u}{a(u-x)} = 0$$

$$\frac{t^2 - uy}{b(y-t)^2} = \frac{u}{a(u-x)}$$

$$\frac{a(u-x)}{b(y-t)} = \frac{u(y-t)}{t^2 - uy}$$

$$\frac{a(u-x)}{b(y-t)} = \frac{-u^2 + tx}{t(u-x)} = \frac{u(y-t)}{t^2 - uy}$$

$$(-u^2 + tx)(t^2 - uy) = tu(u-x)(y-t)$$

$$-t^2u^2 + u^3y + t^3x - tuxy = tu(uy - tu - xy + xt)$$

$$-t^2u^2 + u^3y + t^3x - tuxy = tu^2y - t^2u^2 - tuxy + t^2ux$$

$$u^3y + t^3x = tu^2y + t^2ux$$

$$u^3y - tu^2y = t^2ux - t^3x$$

$$u^2y(u - t) = t^2x(u - t)$$

$$u^2y = t^2x$$

$$\frac{u^2}{t^2} = \frac{x}{y}$$

$$\boxed{\frac{u}{t} = \sqrt{\frac{x}{y}}}$$

Multiply both sides by $a(u-x)$ and divide by t. Note that $\frac{(u-x)}{(u-x)^2} = \frac{1}{(u-x)}$.

Make the first term and the last term have a common denominator.

Distribute the minus sign: $-(ty - uy) = -ty - (-uy) = -ty + uy$. Subtract $\frac{u}{a(u-x)}$ from both sides and multiply both sides by -1.

Multiply by $a(y-t)$ and divide by $t^2 - uy$. Note that $\frac{(y-t)}{(y-t)^2} = \frac{1}{(y-t)}$.

Combine $\frac{a(u-x)}{b(y-t)} = \frac{u(y-t)}{t^2 - uy}$ with $\frac{a(u-x)}{b(y-t)} = \frac{-u^2 + tx}{t(u-x)}$ from earlier. Cross multiply $\frac{-u^2 + tx}{t(u-x)} = \frac{u(y-t)}{t^2 - uy}$.

Factor u^2y on the left and t^2x on the right. Divide both sides by $(u - t)$. Divide both sides by y and by t^2. Square root both sides. Recall that $t > u > 0$.

(B) $\dfrac{t}{b(y-t)} + \dfrac{u}{a(u-x)} - \dfrac{x(t-u)}{a(u-x)^2} = 0$

$\dfrac{t}{b(y-t)} + \dfrac{u}{a(u-x)} - \dfrac{y(t-u)}{b(y-t)^2} = 0$

$\dfrac{x(t-u)}{a(u-x)^2} = \dfrac{y(t-u)}{b(y-t)^2}$

$\dfrac{x}{a(u-x)^2} = \dfrac{y}{b(y-t)^2}$

$bx(y-t)^2 = ay(u-x)^2$

$\sqrt{bx}\,(y-t) = \sqrt{ay}\,(u-x)$

$\dfrac{u}{t} = \sqrt{\dfrac{x}{y}}$

$u = t\sqrt{\dfrac{x}{y}}$

$\sqrt{bx}\,(y-t) = \sqrt{ay}\left(t\sqrt{\dfrac{x}{y}} - x\right)$

$y\sqrt{bx} - t\sqrt{bx} = t\sqrt{ax} - x\sqrt{ay}$

$x\sqrt{ay} + y\sqrt{bx} = t\sqrt{ax} + t\sqrt{bx}$

$\dfrac{x\sqrt{ay} + y\sqrt{bx}}{\sqrt{ax} + \sqrt{bx}} = t$

$\boxed{\dfrac{\sqrt{axy} + y\sqrt{b}}{\sqrt{a} + \sqrt{b}} = t}$

$u = t\sqrt{\dfrac{x}{y}} = \left(\dfrac{\sqrt{axy} + y\sqrt{b}}{\sqrt{a} + \sqrt{b}}\right)\sqrt{\dfrac{x}{y}}$

$u = \dfrac{x\sqrt{ay} + y\sqrt{bx}}{\sqrt{ay} + \sqrt{by}}$

$\boxed{u = \dfrac{x\sqrt{a} + \sqrt{bxy}}{\sqrt{a} + \sqrt{b}}}$

Since the first two terms of these given equations are identical, the third terms must be identical, too. Set these equal. Divide both sides by $t - u$. Cross multiply. Square root both sides. Since the problem states that $y > t > u > x > 0$, only the positive root is applicable. That is, clearly $y - t$ and $u - x$ are both positive. Use the equation from Part A. Distribute on both sides. Note that

$\sqrt{ay}\sqrt{\dfrac{x}{y}} = \sqrt{\dfrac{ayx}{y}} = \sqrt{ax}.$

Add $x\sqrt{ay}$ and $t\sqrt{bx}$ to both sides. Factor out t and divide by $\sqrt{ax} + \sqrt{bx}$. Factor out $\sqrt{x}$ in the numerator and denominator. Distribute $\sqrt{x}$ and $\sqrt{y}$. Factor out $\sqrt{y}$ in the numerator and denominator.

Solution to Exercise #91

$$\frac{t}{b(y-t)} + \frac{u}{a(u-x)} - \frac{x(t-u)}{a(u-x)^2} = 0$$

$$\frac{t}{b(y-t)} + \frac{u}{a(u-x)} = \frac{x(t-u)}{a(u-x)^2}$$

$$\frac{t}{b(y-t)} + \frac{u}{a(u-x)} - \frac{y(t-u)}{b(y-t)^2} = 0$$

$$\frac{t}{b(y-t)} + \frac{u}{a(u-x)} = \frac{y(t-u)}{b(y-t)^2}$$

$$\frac{t}{b(y-t)} + \frac{u}{a(u-x)} = \frac{x(t-u)}{a(u-x)^2} = \frac{y(t-u)}{b(y-t)^2}$$

$$\sqrt{\frac{x(t-u)}{a(u-x)^2}} = \sqrt{\frac{y(t-u)}{b(y-t)^2}}$$

$$\frac{x(t-u)}{a(u-x)^2} = \sqrt{\frac{x(t-u)}{a(u-x)^2}}\sqrt{\frac{x(t-u)}{a(u-x)^2}} = \sqrt{\frac{x(t-u)}{a(u-x)^2}}\sqrt{\frac{y(t-u)}{b(y-t)^2}}$$

$$\frac{x(t-u)}{a(u-x)^2} = \sqrt{\frac{x(t-u)}{a(u-x)^2}\frac{y(t-u)}{b(y-t)^2}} = \frac{\sqrt{xy}(t-u)}{\sqrt{ab}(u-x)(y-t)}$$

$$\boxed{\frac{t}{b(y-t)} + \frac{u}{a(u-x)} = \frac{\sqrt{xy}(t-u)}{\sqrt{ab}(u-x)(y-t)}}$$

$$u = \frac{x\sqrt{a} + \sqrt{bxy}}{\sqrt{a} + \sqrt{b}} = \frac{\sqrt{x}(\sqrt{ax} + \sqrt{by})}{\sqrt{a} + \sqrt{b}}$$

$$u - x = \frac{\sqrt{x}(\sqrt{ax} + \sqrt{by})}{\sqrt{a} + \sqrt{b}} - x = \frac{\sqrt{x}(\sqrt{ax} + \sqrt{by})}{\sqrt{a} + \sqrt{b}} - \frac{x(\sqrt{a} + \sqrt{b})}{\sqrt{a} + \sqrt{b}}$$

$$\boxed{u - x = \frac{\sqrt{x}(\sqrt{ax} + \sqrt{by}) - x(\sqrt{a} + \sqrt{b})}{\sqrt{a} + \sqrt{b}}}$$

$$t = \frac{\sqrt{axy} + y\sqrt{b}}{\sqrt{a} + \sqrt{b}} = \frac{\sqrt{y}(\sqrt{ax} + \sqrt{by})}{\sqrt{a} + \sqrt{b}}$$

$$y - t = y - \frac{\sqrt{y}(\sqrt{ax} + \sqrt{by})}{\sqrt{a} + \sqrt{b}} = \frac{y(\sqrt{a} + \sqrt{b})}{\sqrt{a} + \sqrt{b}} - \frac{\sqrt{y}(\sqrt{ax} + \sqrt{by})}{\sqrt{a} + \sqrt{b}}$$

$$y - t = \frac{y(\sqrt{a} + \sqrt{b}) - \sqrt{y}(\sqrt{ax} + \sqrt{by})}{\sqrt{a} + \sqrt{b}}$$

$$w = \frac{t - u}{\dfrac{t}{b(y - t)} + \dfrac{u}{a(u - x)}} = \frac{t - u}{\dfrac{\sqrt{xy}(t - u)}{\sqrt{ab}(u - x)(y - t)}} = (t - u) \div \frac{\sqrt{xy}(t - u)}{\sqrt{ab}(u - x)(y - t)}$$

$$w = (t - u)\frac{\sqrt{ab}(u - x)(y - t)}{\sqrt{xy}(t - u)} = \frac{\sqrt{ab}(u - x)(y - t)}{\sqrt{xy}}$$

$$w = \frac{\sqrt{ab}}{\sqrt{xy}}\left[\frac{\sqrt{x}(\sqrt{ax} + \sqrt{by}) - x(\sqrt{a} + \sqrt{b})}{\sqrt{a} + \sqrt{b}}\right]\left[\frac{y(\sqrt{a} + \sqrt{b}) - \sqrt{y}(\sqrt{ax} + \sqrt{by})}{\sqrt{a} + \sqrt{b}}\right]$$

$$w = \frac{\sqrt{ab}}{(\sqrt{a} + \sqrt{b})^2}[(\sqrt{ax} + \sqrt{by}) - \sqrt{x}(\sqrt{a} + \sqrt{b})][\sqrt{y}(\sqrt{a} + \sqrt{b}) - (\sqrt{ax} + \sqrt{by})]$$

The first brackets were divided by $\sqrt{x}$. The second brackets were divided by $\sqrt{y}$.

$$[(\sqrt{ax} + \sqrt{by}) - \sqrt{x}(\sqrt{a} + \sqrt{b})][\sqrt{y}(\sqrt{a} + \sqrt{b}) - (\sqrt{ax} + \sqrt{by})]$$

$$= (\sqrt{ax} + \sqrt{by})\sqrt{y}(\sqrt{a} + \sqrt{b}) - (\sqrt{ax} + \sqrt{by})(\sqrt{ax} + \sqrt{by})$$

$$- \sqrt{x}(\sqrt{a} + \sqrt{b})\sqrt{y}(\sqrt{a} + \sqrt{b}) + \sqrt{x}(\sqrt{a} + \sqrt{b})(\sqrt{ax} + \sqrt{by})$$

$$= a\sqrt{xy} + \sqrt{abxy} + y\sqrt{ab} + by - ax - \sqrt{abxy} - \sqrt{abxy} - by - a\sqrt{xy} - \sqrt{abxy}$$

$$- \sqrt{abxy} - b\sqrt{xy} + ax + \sqrt{abxy} + x\sqrt{ab} + b\sqrt{xy}$$

$$= y\sqrt{ab} - 2\sqrt{abxy} + x\sqrt{ab} = \sqrt{ab}(y - 2\sqrt{xy} + x) = \sqrt{ab}(\sqrt{y} - \sqrt{x})^2$$

$$w = \frac{\sqrt{ab}}{(\sqrt{a} + \sqrt{b})^2}\sqrt{ab}(\sqrt{y} - \sqrt{x})^2$$

$$w = ab\left(\frac{\sqrt{y} - \sqrt{x}}{\sqrt{a} + \sqrt{b}}\right)^2$$

Note: The algebra in Exercises 90-91 resembles the calculation of the power delivered by an endoreversible Carnot-like heat engine in thermodynamics, where the formulas are:

$$\frac{T_{out}}{T_{in}} = \sqrt{\frac{T_c}{T_h}}$$

$$T_{in} = \frac{\sqrt{\sigma_c T_c T_h} + T_h\sqrt{\sigma_h}}{\sqrt{\sigma_c} + \sqrt{\sigma_h}}$$

$$T_{out} = \frac{T_c\sqrt{\sigma_c} + \sqrt{\sigma_h} T_c T_h}{\sqrt{\sigma_c} + \sqrt{\sigma_h}}$$

$$P_{net} = \frac{T_{in} - T_{out}}{\dfrac{T_{in}}{\sigma_h(T_h - T_{in})} + \dfrac{T_{out}}{\sigma_c(T_{out} - T_c)}}$$

$$P_{net} = \sigma_c\sigma_h\left(\frac{\sqrt{T_h} - \sqrt{T_c}}{\sqrt{\sigma_c} + \sqrt{\sigma_h}}\right)^2$$

Solution to Exercise #92

(A) $2ax\sqrt{y} + b\sqrt{y} = 0$

$$\sqrt{y}(2ax + b) = 0$$

$$2ax + b = 0$$

$$2ax = -b$$

$$\boxed{x = -\frac{b}{2a}}$$

(B) $ax^2 + ay + bx + c = 0$

$$a\left(-\frac{b}{2a}\right)^2 + ay + b\left(-\frac{b}{2a}\right) + c = 0$$

$$a\left(\frac{b^2}{4a^2}\right) + ay - \frac{b^2}{2a} + c = 0$$

$$\frac{b^2}{4a} + ay - \frac{b^2}{2a} + c = 0$$

$$ay - \frac{b^2}{4a} + c = 0$$

$$ay = \frac{b^2}{4a} - c$$

$$\boxed{y} = \frac{b^2}{4a^2} - \frac{c}{a} = \frac{b^2}{4a^2} - \frac{4ac}{4a^2} = \boxed{\frac{b^2 - 4ac}{4a^2}}$$

(C) $x \pm \sqrt{y} = -\frac{b}{2a} \pm \sqrt{\frac{b^2 - 4ac}{4a^2}}$

$$= -\frac{b}{2a} \pm \frac{\sqrt{b^2 - 4ac}}{2a}$$

$$= \frac{-b \pm \sqrt{b^2 - 4ac}}{2a}$$

$$a\left(x \pm \sqrt{y}\right)^2 + b\left(x \pm \sqrt{y}\right) + c$$

$$= ax^2 \pm 2ax\sqrt{y} + ay + bx$$
$$\pm b\sqrt{y} + c$$

$$= ax^2 + ay + bx + c \pm \left(2ax\sqrt{y} + b\sqrt{y}\right)$$

$$= 0 \pm 0 = 0$$

$$w = x \pm \sqrt{y}$$

$$aw^2 + bw + c = 0$$

$$w = \frac{-b \pm \sqrt{b^2 - 4ac}}{2a}$$

Note: This problem is modeled after one method of deriving the quadratic formula.

Solution to Exercise #93	
(A) $\dfrac{3}{w} > \dfrac{1}{w-2}$	$\dfrac{4}{y} < \dfrac{3}{y-6}$

(A) $\dfrac{3}{w} > \dfrac{1}{w-2}$

$3w - 6 > w$ or $0 < w < 2$

$2w > 6$ or $0 < w < 2$

$\boxed{w > 3}$ or $\boxed{0 < w < 2}$

$\dfrac{8}{x} > \dfrac{5}{x+3}$

$8x + 24 > 5x$ or $x > 0$

$3x > -24$ or $x > 0$

$\boxed{-8 < x < -3}$ or $\boxed{x > 0}$

$\dfrac{4}{y} < \dfrac{3}{y-6}$

$4y - 24 < 3y$ or $y < 0$

$\boxed{6 < y < 24}$ or $\boxed{y < 0}$

$\dfrac{9}{z} < \dfrac{7}{z+4}$

$-4 < z < 0$ or $9z + 36 < 7z$

$-4 < z < 0$ or $2z < -36$

$\boxed{-4 < z < 0}$ or $\boxed{z < -18}$

(B) The direction of the inequality reverses when multiplying both sides of the inequality by a negative number, but remains unchanged when multiplying both sides of the inequality by a positive number. When cross multiplying, the inequality is multiplied by two values. For example, $\frac{8}{x} > \frac{5}{x+3}$ involves multiplying both sides by both x and $x + 3$. If x and $x + 3$ have the same sign (both positive or both negative), then the direction of the inequality remains unchanged, but if x and $x + 3$ have opposite signs, then the direction of the inequality reverses. In this example, for values of x between -3 and 0, the quantity x is negative whereas $x + 3$ is positive, so the direction of the inequality reverses for this interval of values of x. This means that the interval -3 and 0 must be considered separately from other values of x. In this example, the algebra leads to $x > -8$ when x and $x + 3$ are both negative (which requires $x < -3$), the given inequality is not satisfied if x and $x + 3$ have opposite signs, and the given inequality is satisfied when $x > 0$. This example shows that it is necessary to consider multiple intervals when cross multiplying to isolate a variable in an inequality.

(C) $\dfrac{3}{w} > \dfrac{1}{w-2}$	$\dfrac{4}{y} < \dfrac{3}{y-6}$
$w = 3.1$ is > 3	$y = 6.1$ lies in $6 < y < 24$
$\dfrac{3}{3.1} \approx 0.968$ is $> \dfrac{1}{3.1-2} \approx 0.909$ ✔	$\dfrac{4}{6.1} \approx 0.656$ is $< \dfrac{3}{6.1-6} = 30$ ✔

$$w = 2.9 \text{ is NOT} > 3$$

$$\frac{3}{2.9} \approx 1.034 \text{ is NOT} > \frac{1}{2.9 - 2} \approx 1.111 \quad \checkmark$$

$$w = 1.9 \text{ lies in } 0 < w < 2$$

$$\frac{3}{1.9} \approx 1.579 \text{ is} > \frac{1}{1.9 - 2} = -10 \quad \checkmark$$

$$w = 2.1 \text{ does NOT lie in } 0 < w < 2$$

$$\frac{3}{2.1} \approx 1.429 \text{ is NOT} > \frac{1}{2.1 - 2} = 10 \quad \checkmark$$

$$w = 0.1 \text{ lies in } 0 < w < 2$$

$$\frac{3}{0.1} = 30 \text{ is} > \frac{1}{0.1 - 2} \approx -0.526 \quad \checkmark$$

$$w = -0.1 \text{ does NOT lie in } 0 < w < 2$$

$$\frac{3}{-0.1} = -30 \text{ is NOT} > \frac{1}{-0.1 - 2} \approx -0.476 \quad \checkmark$$

$$\frac{8}{x} > \frac{5}{x + 3}$$

$$x = -3.1 \text{ lies in } -8 < x < -3$$

$$\frac{8}{-3.1} \approx -2.581 \text{ is} > \frac{5}{-3.1 + 3} = -50 \quad \checkmark$$

$$x = -2.9 \text{ does NOT lie in } -8 < x < -3$$

$$\frac{8}{-2.9} \approx -2.759 \text{ is NOT} > \frac{5}{-2.9 + 3}$$

$$= 50 \quad \checkmark$$

$$x = -7.9 \text{ lies in } -8 < x < -3$$

$$\frac{8}{-7.9} \approx -1.013 \text{ is} > \frac{5}{-7.9 + 3} \approx -1.020 \quad \checkmark$$

$$x = -8.1 \text{ does NOT lie in } -8 < x < -3$$

$$\frac{8}{-8.1} \approx -0.988 \text{ is NOT} > \frac{5}{-8.1 + 3} \approx -0.980 \quad \checkmark$$

$$x = 0.1 \text{ is} > 0$$

$$\frac{8}{0.1} = 80 \text{ is} > \frac{5}{0.1 + 3} \approx 1.613 \quad \checkmark$$

$$x = -0.1 \text{ is NOT} > 0$$

$$\frac{8}{-0.1} = -80 \text{ is NOT} > \frac{5}{-0.1 + 3} \approx 1.724 \quad \checkmark$$

$$y = 5.9 \text{ does NOT lie in } 6 < y < 24$$

$$\frac{4}{5.9} \approx 0.678 \text{ is NOT} < \frac{3}{5.9 - 6} = -30 \quad \checkmark$$

$$y = 23.9 \text{ lies in } 6 < y < 24$$

$$\frac{4}{23.9} \approx 0.1674 \text{ is} < \frac{3}{23.9 - 6} \approx 0.1676 \quad \checkmark$$

$$y = 24.1 \text{ does NOT lie in } 6 < y < 24$$

$$\frac{4}{24.1} \approx 0.1660 \text{ is NOT} < \frac{3}{24.1 - 6} \approx 0.1657 \quad \checkmark$$

$$y = -0.1 \text{ is} < 0$$

$$\frac{4}{-0.1} = -40 \text{ is} < \frac{3}{-0.1 - 6} \approx -0.492 \quad \checkmark$$

$$y = 0.1 \text{ is NOT} < 0$$

$$\frac{4}{0.1} = 40 \text{ is NOT} < \frac{3}{0.1 - 6} \approx -0.508 \quad \checkmark$$

$$\frac{9}{z} < \frac{7}{z + 4}$$

$$z = -3.9 \text{ lies in } -4 < z < 0$$

$$\frac{9}{-3.9} \approx 2.308 \text{ is} < \frac{7}{-3.9 + 4} = 70 \quad \checkmark$$

$$z = -4.1 \text{ does NOT lie in } -4 < z < 0$$

$$\frac{9}{-4.1} \approx -2.195 \text{ is NOT} < \frac{7}{-4.1 + 4} = -70 \quad \checkmark$$

$$z = -0.1 \text{ lies in } -4 < z < 0$$

$$\frac{9}{-0.1} = -90 \text{ is} < \frac{7}{-0.1 + 4} \approx 1.795 \quad \checkmark$$

$$z = 0.1 \text{ does NOT lie in } -4 < z < 0$$

$$\frac{9}{0.1} = 90 \text{ is NOT} < \frac{7}{0.1 + 4} \approx 1.707 \quad \checkmark$$

$$z = -18.1 \text{ is} < -18$$

$$\frac{9}{-18.1} \approx -0.4972 \text{ is} < \frac{7}{-18.1 + 4} \approx -0.4965 \quad \checkmark$$

$$z = -17.9 \text{ is NOT} < -18$$

$$\frac{9}{-17.9} \approx -0.5028 \text{ is NOT} < \frac{7}{-17.9 + 4}$$

$$\approx -0.5036 \quad \checkmark$$

Solution to Exercise #94	
(A) $x^2 + y^2 = 25$	(B) $x^2 + y^2 = 25$
$(-3)^2 + y^2 = 25$	$3^4 + 4^2 = 9 + 16 = 25$ ✓
$9 + y^2 = 25$	$y = -\dfrac{3}{4}x + \dfrac{25}{4}$
$y^2 = 25 - 9 = 16$	
$y = \sqrt{16} = 4$	$4 = -\dfrac{3}{4}3 + \dfrac{25}{4} = -\dfrac{9}{4} + \dfrac{25}{4} = \dfrac{16}{4} = 4$ ✓
$m_\perp = \dfrac{4-0}{3-0} = \dfrac{4}{3}$	(C) $y = m_\perp x + b_\perp$
$m = -\dfrac{1}{m_\perp} = -\dfrac{3}{4}$	$4 = \dfrac{4}{3}(3) + b_\perp$
$y = mx + b$	$4 = 4 + b_\perp$
$4 = -\dfrac{3}{4}(3) + b$	$0 = b_\perp$
$4 = -\dfrac{9}{4} + b$	$y = \dfrac{4}{3}x$
$\dfrac{9}{4} + 4 = \dfrac{9}{4} + \dfrac{16}{4} = \dfrac{25}{4} = b$	$y = -\dfrac{3}{4}x + \dfrac{25}{4}$
$\boxed{y = -\dfrac{3}{4}x + \dfrac{25}{4}}$	$m = -\dfrac{1}{m_\perp} = -\dfrac{3}{4}$ ✓

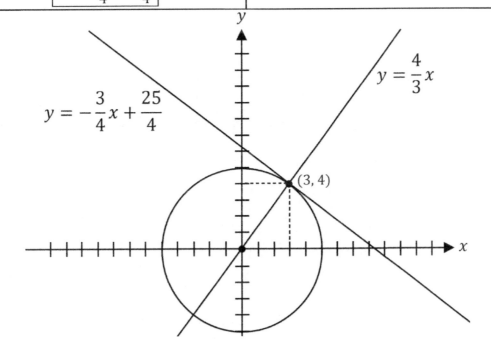

$$y = -\frac{3}{4}x + \frac{25}{4}$$

$$y = \frac{4}{3}x$$

$(3, 4)$

Solution to Exercise #95

(A) $\dfrac{|x-y|}{x} = 0.1$

$$|x-y| = 0.1x$$

$$x - y = \pm 0.1x$$

$$x = y \pm 0.1x$$

$$x \mp 0.1x = y$$

$$3x + 6y = 8.4$$

$$3x + 6(x \mp 0.1x) = 8.4$$

$$3x + 6x \mp 0.6x = 8.4$$

$$9x \mp 0.6x = 8.4$$

$$(9 \mp 0.6)x = 8.4$$

$$x = \frac{8.4}{9 \mp 0.6} = \frac{8.4}{9 \mp 0.6}\left(\frac{5}{5}\right) = \frac{42}{45 \mp 3}$$

$$x = \frac{42}{45 - 3} \quad \text{or} \quad x = \frac{42}{45 + 3}$$

$$x = \frac{42}{42} = \boxed{1} \quad \text{or} \quad x = \frac{42}{48} = \boxed{\frac{7}{8}} = \boxed{0.875}$$

$$y = x \mp 0.1x = (1 \mp 0.1)x$$

$$y = (1 - 0.1)(1) \quad \text{or} \quad y = (1 + 0.1)\left(\frac{7}{8}\right)$$

$$y = (0.9)(1) \quad \text{or} \quad y = (1.1)\left(\frac{7}{8}\right)$$

$$y = \left(\frac{9}{10}\right)(1) = \boxed{\frac{9}{10}} = \boxed{0.9} \quad \text{or} \quad y = \left(\frac{11}{10}\right)\left(\frac{7}{8}\right) = \boxed{\frac{77}{80}} = \boxed{0.9625}$$

(B) $\dfrac{|x-y|}{x} = 0.1$

$$\frac{|1 - 0.9|}{1} = \frac{|0.1|}{1} = \frac{0.1}{1} = 0.1 \quad \checkmark$$

$$\frac{|0.875 - 0.9625|}{0.875} = \frac{|-0.0875|}{0.875} = \frac{0.0875}{0.875} = 0.1 \quad \checkmark$$

$$3x + 6y = 8.4$$

$$3(1) + 6(0.9) = 3 + 5.4 = 8.4 \quad \checkmark$$

$$3(0.875) + 6(0.9625) = 2.625 + 5.775 = 8.4 \quad \checkmark$$

Note: The equation $\frac{|x-y|}{x} = 0.1$ is used in science and engineering to calculate percent error when a measured value (y) is compared to an expected value (x).

Solution to Exercise #96

(A) $y = ax^3 + bx^2 + cx + d$

$$y = a\left(t - \frac{b}{3a}\right)^3 + b\left(t - \frac{b}{3a}\right)^2 + c\left(t - \frac{b}{3a}\right) + d$$

$$y = a\left(t^3 - \frac{2bt^2}{3a} + \frac{b^2t}{9a^2} - \frac{bt^2}{3a} + \frac{2b^2t}{9a^2} - \frac{b^3}{27a^3}\right) + b\left(t^2 - \frac{2bt}{3a} + \frac{b^2}{9a^2}\right) + ct - \frac{bc}{3a} + d$$

$$y = a\left(t^3 - \frac{bt^2}{a} + \frac{b^2t}{3a^2} - \frac{b^3}{27a^3}\right) + bt^2 - \frac{2b^2t}{3a} + \frac{b^3}{9a^2} + ct - \frac{bc}{3a} + d$$

$$y = at^3 - bt^2 + \frac{b^2t}{3a} - \frac{b^3}{27a^2} + bt^2 - \frac{2b^2t}{3a} + \frac{b^3}{9a^2} + ct - \frac{bc}{3a} + d$$

$$y = at^3 + \left(\frac{b^2}{3a} - \frac{2b^2}{3a} + c\right)t + \left(-\frac{b^3}{27a^2} + \frac{b^3}{9a^2} - \frac{bc}{3a} + d\right)$$

$$y = at^3 + pt + q$$

$$p = \frac{b^2}{3a} - \frac{2b^2}{3a} + c \quad , \quad q = -\frac{b^3}{27a^2} + \frac{b^3}{9a^2} - \frac{bc}{3a} + d$$

$$p = -\frac{b^2}{3a} + c \quad , \quad q = -\frac{b^3}{27a^2} + \frac{3b^3}{27a^2} - \frac{bc}{3a} + d$$

$$\boxed{p = c - \frac{b^2}{3a}} \quad , \quad \boxed{q = \frac{2b^3}{27a^2} - \frac{bc}{3a} + d}$$

(B) $y = at^3 + pt + q$

$$\frac{z}{\sqrt{a}} + q = a\left(\frac{u}{\sqrt{a}}\right)^3 + p\left(\frac{u}{\sqrt{a}}\right) + q$$

$$\frac{z}{\sqrt{a}} = a\left(\frac{u^3}{a\sqrt{a}}\right) + \frac{pu}{\sqrt{a}}$$

$$\frac{z}{\sqrt{a}} = \frac{u^3}{\sqrt{a}} + \frac{pu}{\sqrt{a}}$$

$$\boxed{z = u^3 + pu}$$

Note: The algebra featured in this problem is part of the classic strategy for solving a cubic equation (which is somewhat more complicated than the quadratic).

Solution to Exercise #97

Method 1: Brute Force	Method 2: Completing the Square

Method 1: Brute Force

$$(A)\ z = \sqrt{\frac{1}{2} - \frac{\sqrt{3}}{4}} = \sqrt{x} - \sqrt{y}$$

$$z^2 = \frac{1}{2} - \frac{\sqrt{3}}{4} = \left(\sqrt{x} - \sqrt{y}\right)^2$$

$$\frac{1}{2} - \frac{\sqrt{3}}{4} = x - 2\sqrt{xy} + y$$

$$x + y = \frac{1}{2}\ ,\quad \frac{\sqrt{3}}{4} = 2\sqrt{xy}$$

$$\frac{3}{16} = 4xy$$

$$\frac{3}{64x} = y$$

$$x + \frac{3}{64x} = \frac{1}{2}$$

$$64x^2 + 3 = 32x$$

$$64x^2 - 32x + 3 = 0$$

$$(8x - 3)(8x - 1) = 0$$

$$8x - 3 = 0 \quad \text{or} \quad 8x - 1 = 0$$

$$8x = 3 \quad \text{or} \quad 8x = 1$$

$$x = \boxed{\frac{3}{8}} = \boxed{0.375} \quad \text{only}$$

Since $z > 0$, $\sqrt{x} - \sqrt{y} > 0$, so that $x > y$.

The problem with $8x = 1$ is that $x = \frac{1}{8}$

leads to $y = \frac{3}{8}$, which is greater than x.

$$y = \frac{3}{64x}$$

$$y = \frac{3}{64\left(\frac{3}{8}\right)} = \frac{3}{24} = \boxed{\frac{1}{8}} = \boxed{0.125}$$

Method 2: Completing the Square

$$(A)\ \frac{1}{2} - \frac{\sqrt{3}}{4} = \frac{2}{4} - \frac{\sqrt{3}}{4} = \frac{2 - \sqrt{3}}{4}$$

$$2 - \sqrt{3} = \frac{3}{2} - \sqrt{3} + \frac{1}{2} = \frac{6}{4} - \frac{4\sqrt{3}}{4} + \frac{2}{4}$$

$$= \frac{6 - 4\sqrt{3} + 2}{4}$$

$$6 - 4\sqrt{3} + 2 = 6 - 2(2)\sqrt{3} + 2$$

$$= 6 - 2\sqrt{(4)(3)} + 2$$

Note: $(4)(3) = (6)(2)$.

$$6 - 4\sqrt{3} + 2 = 6 - 2\sqrt{6}\sqrt{2} + 2$$

$$= \left(\sqrt{6} - \sqrt{2}\right)^2$$

$$z = \sqrt{\frac{1}{2} - \frac{\sqrt{3}}{4}} = \sqrt{\frac{2 - \sqrt{3}}{4}} = \frac{1}{2}\sqrt{2 - \sqrt{3}}$$

$$z = \frac{1}{2}\sqrt{\frac{6 - 4\sqrt{3} + 2}{4}} = \frac{1}{4}\sqrt{6 - 4\sqrt{3} + 2}$$

$$z = \frac{1}{4}\sqrt{\left(\sqrt{6} - \sqrt{2}\right)^2} = \frac{\sqrt{6} - \sqrt{2}}{4}$$

$$z = \sqrt{x} - \sqrt{y}$$

$$\sqrt{x} = \frac{\sqrt{6}}{4}$$

$$x = \left(\frac{\sqrt{6}}{4}\right)^2 = \frac{6}{16} = \boxed{\frac{3}{8}} = \boxed{0.375}$$

$$\sqrt{y} = \frac{\sqrt{2}}{4}$$

$$y = \left(\frac{\sqrt{2}}{4}\right)^2 = \frac{2}{16} = \boxed{\frac{1}{8}} = \boxed{0.125}$$

(B) $z = \sqrt{x} - \sqrt{y} = \sqrt{\dfrac{3}{8}} - \sqrt{\dfrac{1}{8}}$

$z = \dfrac{\sqrt{3}}{\sqrt{8}} - \dfrac{\sqrt{1}}{\sqrt{8}} = \dfrac{\sqrt{3}}{\sqrt{8}}\left(\dfrac{\sqrt{2}}{\sqrt{2}}\right) - \dfrac{\sqrt{1}}{\sqrt{8}}\left(\dfrac{\sqrt{8}}{\sqrt{8}}\right)$

$z = \dfrac{\sqrt{6}}{\sqrt{16}} - \dfrac{\sqrt{8}}{8} = \dfrac{\sqrt{6}}{4} - \dfrac{\sqrt{(4)(2)}}{8}$

$z = \dfrac{\sqrt{6}}{4} - \dfrac{2\sqrt{2}}{8} = \boxed{\dfrac{\sqrt{6}}{4} - \dfrac{\sqrt{2}}{4}} = \boxed{\dfrac{\sqrt{6} - \sqrt{2}}{4}}$

Note: This nested square root calculation arises in trigonometry when applying the half-angle identity to compute the sine of 15° without a calculator. Of course, knowledge of trigonometry is NOT needed to solve this algebra problem.

(C) $z = \dfrac{\sqrt{6} - \sqrt{2}}{4} = \sqrt{\dfrac{1}{2} - \dfrac{\sqrt{3}}{4}}$

$z^2 = \left(\dfrac{\sqrt{6} - \sqrt{2}}{4}\right)^2 = \dfrac{1}{2} - \dfrac{\sqrt{3}}{4}$

$\dfrac{6 - 2\sqrt{6}\sqrt{2} + 2}{16} = \dfrac{1}{2} - \dfrac{\sqrt{3}}{4}$

$\dfrac{8 - 2\sqrt{12}}{16} = \dfrac{1}{2} - \dfrac{\sqrt{3}}{4}$

$\dfrac{8}{16} - \dfrac{2\sqrt{12}}{16} = \dfrac{1}{2} - \dfrac{\sqrt{3}}{4}$

$\dfrac{1}{2} - \dfrac{\sqrt{12}}{8} = \dfrac{1}{2} - \dfrac{\sqrt{3}}{4}$

$\dfrac{1}{2} - \dfrac{\sqrt{(4)(3)}}{8} = \dfrac{1}{2} - \dfrac{\sqrt{3}}{4}$

$\dfrac{1}{2} - \dfrac{2\sqrt{3}}{8} = \dfrac{1}{2} - \dfrac{\sqrt{3}}{4}$

$\dfrac{1}{2} - \dfrac{\sqrt{3}}{4} = \dfrac{1}{2} - \dfrac{\sqrt{3}}{4}$ ✓

Solution to Exercise #98	
$x + kc = y + \sqrt{z^2 + k^2 c^2}$	Subtract y from both sides.
$x + kc - y = \sqrt{z^2 + k^2 c^2}$	Square both sides.
$x^2 + 2kcx + k^2 c^2 - 2kcy + y^2 - 2xy = z^2 + k^2 c^2$	Subtract $k^2 c^2$. It cancels.
$x^2 + 2kcx - 2kcy + y^2 - 2xy = z^2$	
$x = yq + zs$	Subtract yq.
$x - yq = zs$	Square both sides.
$x^2 - 2xyq + y^2 q^2 = z^2 s^2$	
$0 = yr - zt$	Add zt to both sides.
$yr = zt$	Square both sides.
$y^2 r^2 = z^2 t^2$	Add two equations.
$x^2 - 2xyq + y^2 q^2 + y^2 r^2 = z^2 s^2 + z^2 t^2$	Factor y^2 and z^2.

$$x^2 - 2xyq + y^2(q^2 + r^2) = z^2(s^2 + t^2)$$

$q^2 + r^2 = s^2 + t^2 = 1$	Plug these equations into the previous equation.
$x^2 - 2xyq + y^2 = z^2$	
$-x^2 + 2xyq - y^2 = -z^2$	Multiply both sides by -1.
$x^2 + 2kcx - 2kcy + y^2 - 2xy = z^2$	Recall this equation.
$2kcx - 2kcy + 2xyq - 2xy = 0$	Add two equations.
$2kcx - 2kcy = 2xy - 2xyq$	Add $2xy$. Subtract $2xyq$.
$2kc(x - y) = 2xy(1 - q)$	Factor $2kc$ and $2xy$.
$x - y = \dfrac{xy}{kc}(1 - q)$	Divide both sides by kc.
$\dfrac{x}{xy} - \dfrac{y}{xy} = \dfrac{1 - q}{kc}$	Divide both sides by xy.
$\boxed{\dfrac{1}{y} - \dfrac{1}{x} = \dfrac{1 - q}{kc}}$	

Note: This problem is modeled after a classic Compton scattering problem in physics, where the formulas are:

$$p_{\gamma i} = p_{\gamma f} \cos\theta + p_{ef} \cos\varphi$$

$$0 = p_{\gamma f} \sin\theta + p_{ef} \sin\varphi$$

$$p_{\gamma i}c + m_e c^2 = p_{\gamma f}c + \sqrt{p_{ef}^2 c^2 + m_e^2 c^4}$$

$$\frac{1}{p_{\gamma f}} - \frac{1}{p_{\gamma i}} = \frac{1 - \cos\theta}{m_e c}$$

Since $p_\gamma = \frac{h}{\lambda}$, the last formula above leads to the Compton effect formula:

$$\frac{1}{\lambda_f} - \frac{1}{\lambda_i} = \frac{h}{m_e c}(1 - \cos\theta)$$

Of course, it is NOT necessary to know any physics (or trig) to solve this algebra problem.

Solution to Exercise #99

(A) $\dfrac{1}{x} + \dfrac{1}{y} = \dfrac{1}{3}$

$\dfrac{1}{y} = \dfrac{1}{3} - \dfrac{1}{x} = \dfrac{x}{3x} - \dfrac{3}{3x} = \dfrac{x-3}{3x}$

$y = \dfrac{3x}{x-3}$

$\dfrac{1}{w} + \dfrac{1}{z} = \dfrac{1}{4}$

$\dfrac{1}{z} = \dfrac{1}{4} - \dfrac{1}{w} = \dfrac{w}{4w} - \dfrac{4}{4w} = \dfrac{w-4}{4w}$

$z = \dfrac{4w}{w-4}$

$w + y = 18$

$z - x = 8$

$w + \dfrac{3x}{x-3} = 18$

$\dfrac{4w}{w-4} - x = 8$

$w(x-3) + 3x = 18(x-3)$

$4w - x(w-4) = 8(w-4)$

$wx - 3w + 3x = 18x - 54$

$4w - xw + 4x = 8w - 32$

$wx - 3w = 15x - 54$

$-xw + 4x = 4w - 32$

$-3w + 4x = 15x - 54 + 4w - 32$

$86 = 11x + 7w$

$\dfrac{86 - 11x}{7} = w$

$-xw + 4x = 4w - 32$

$4x = 4w + xw - 32$

$4x = (4+x)w - 32$

(B) $\dfrac{1}{x} + \dfrac{1}{y} = \dfrac{1}{3}$

$-\dfrac{11}{30} + \dfrac{7}{10} = -\dfrac{11}{30} + \dfrac{21}{30} = \dfrac{10}{30} = \dfrac{1}{3}$ ✓

$\dfrac{1}{4} + \dfrac{1}{12} = \dfrac{3}{12} + \dfrac{1}{12} = \dfrac{4}{12} = \dfrac{1}{3}$ ✓

$\dfrac{1}{w} + \dfrac{1}{z} = \dfrac{1}{4}$

$\dfrac{7}{116} + \dfrac{11}{58} = \dfrac{7}{116} + \dfrac{22}{116} = \dfrac{29}{116} = \dfrac{1}{4}$ ✓

$\dfrac{1}{6} + \dfrac{1}{12} = \dfrac{2}{12} + \dfrac{1}{12} = \dfrac{3}{12} = \dfrac{1}{4}$ ✓

$w + y = 18$

$\dfrac{116}{7} + \dfrac{10}{7} = \dfrac{126}{7} = 18$ ✓

$6 + 12 = 18$ ✓

$z - x = 8$

$\dfrac{58}{11} - \left(-\dfrac{30}{11}\right) = \dfrac{58}{11} + \dfrac{30}{11} = \dfrac{88}{11} = 8$ ✓

$12 - 4 = 8$ ✓

$$4x = (4 + x)\frac{86 - 11x}{7} - 32$$

$$28x = (4 + x)(86 - 11x) - 224$$

$$28x = 344 - 44x + 86x - 11x^2 - 224$$

$$0 = -11x^2 + 14x + 120$$

$$0 = (11x + 30)(-x + 4)$$

$$11x + 30 = 0 \quad \text{or} \quad -x + 4 = 0$$

$$11x = -30 \quad \text{or} \quad 4 = x$$

$$x = \boxed{-\frac{30}{11}} \quad \text{or} \quad \boxed{4} = x$$

$$z - x = 8$$

$$z = x + 8$$

$$z = -\frac{30}{11} + 8 \quad \text{or} \quad z = 4 + 8$$

$$z = -\frac{30}{11} + \frac{88}{11} = \boxed{\frac{58}{11}} \quad \text{or} \quad z = \boxed{12}$$

$$w = \frac{86 - 11x}{7}$$

$$w = \frac{86 - 11\left(-\frac{30}{11}\right)}{7}$$

$$\text{or} \quad w = \frac{86 - 11(4)}{7}$$

$$w = \frac{86 + 30}{7} = \boxed{\frac{116}{7}}$$

$$\text{or} \quad w = \frac{86 - 44}{7} = \frac{42}{7} = \boxed{6}$$

$$w + y = 18$$

$$y = 18 - w$$

$$y = 18 - \frac{116}{7} \quad \text{or} \quad y = 18 - 6$$

$$y = \frac{126}{7} - \frac{116}{7} = \boxed{\frac{10}{7}} \quad \text{or} \quad y = \boxed{12}$$

Solution to Exercise #100

$$(A)\ z = \frac{wt}{2}$$

$$z = \sqrt{\frac{w^2 t^2}{4}} = \sqrt{\frac{4w^2 t^2}{16}}$$

$$t^2 = u^2 - y^2$$

$$z = \sqrt{\frac{4w^2(u^2 - y^2)}{16}} = \sqrt{\frac{4w^2 u^2 - 4w^2 y^2}{16}}$$

$$2wy = u^2 + w^2 - x^2$$

$$4w^2 y^2 = (u^2 + w^2 - x^2)^2$$

$$\boxed{z = \sqrt{\frac{4u^2 w^2 - (u^2 + w^2 - x^2)^2}{16}}}$$

$$(B)\ a = 2uw \quad,\quad b = u^2 + w^2 - x^2$$

$$z = \sqrt{\frac{a^2 - b^2}{16}} = \sqrt{\frac{(a+b)(a-b)}{16}}$$

$$-b = -u^2 - w^2 + x^2$$

$$\boxed{z = \sqrt{\frac{(2uw + u^2 + w^2 - x^2)(2uw - u^2 - w^2 + x^2)}{16}}}$$

$$(C)\ (u+w)^2 = u^2 + 2uw + w^2$$

$$(u-w)^2 = u^2 - 2uw + w^2$$

$$-(u-w)^2 = -u^2 + 2uw - w^2$$

$$\boxed{z = \sqrt{\frac{[(u+w)^2 - x^2][x^2 - (u-w)^2]}{16}}}$$

$$(D)\ c = u + w$$

$$d = u - w$$

$$c^2 - x^2 = (c+x)(c-x)$$

$$x^2 - d^2 = (x+d)(x-d)$$

$$x - d = x - (u - w) = x - u + w$$

$$z = \sqrt{\frac{(u + w + x)(u + w - x)(u + x - w)(w + x - u)}{16}}$$

$$(E)\ 2s = u + w + x$$

$$s = \frac{u + w + x}{2}$$

$$s - u = \frac{u + w + x}{2} - \frac{2u}{2} = \frac{w + x - u}{2}$$

$$s - w = \frac{u + w + x}{2} - \frac{2w}{2} = \frac{u + x - w}{2}$$

$$s - x = \frac{u + w + x}{2} - \frac{2x}{2} = \frac{u + w - x}{2}$$

$$z = \sqrt{\left(\frac{u + w + x}{2}\right)\left(\frac{u + w - x}{2}\right)\left(\frac{u + x - w}{2}\right)\left(\frac{w + x - u}{2}\right)}$$

$$z = \sqrt{s(s - x)(s - w)(s - u)}$$

Note: The algebra for this problem appears in one method of deriving Heron's formula in geometry, which provides a formula for the area of a triangle in terms of the lengths of the three sides and the semiperimeter (whereas the more common formula for the area of a triangle uses the base and height).

Solution to Exercise #101	
(A) $z = (x + iy)(t + iu)$ $z = tx + iux + ity + i^2uy$ $z = tx + iux + ity - uy$ $z = (tx - uy) + i(ux + ty)$ $z = (t + u)x(1 + i) - pu + iqt$ $z = tx + itx + ux + iux - pu + iqt$ $z = (tx + ux - pu) + i(tx + ux + qt)$ Compare $z = (tx - uy) + i(ux + ty)$ with $z =$ $(tx + ux - pu) + i(tx + ux + qt)$. Set the real	(B) Given $t, u, x,$ and y, the formula $z = tx + iux + ity - uy$ involves four multiplications ($tx, ux, ty,$ and uy), whereas the formula below involves only three multiplications: $(t + u)x, (x + y)u,$ and $(y - x)t$. $z = (t + u)x(1 + i) - (x + y)u$ $+ i(y - x)t$

part of each expression equal to get the first equation below. Set the imaginary part of each expression equal to get the second equation below.

$$tx - uy = tx + ux - pu$$
$$ux + ty = tx + ux + qt$$

Subtract tx from both sides of the first equation and ux from the second equation.

$$-uy = ux - pu$$
$$ty = tx + qt$$

Divide by u on both sides of the first equation and divide by t in the second equation.

$$-y = x - p$$
$$y = x + q$$

Add p to both sides of the first equation.

$$p - y = x$$

Now add y to both sides.

$$\boxed{p = x + y}$$

Isolate q in the second equation.

$$\boxed{y - x = q}$$

Check:

$$z = (t + u)x(1 + i) - pu + iqt$$
$$z = (t + u)x(1 + i) - (x + y)u + i(y - x)t$$
$$z = tx + itx + ux + iux - ux - uy + ity - itx$$
$$z = tx + iux - uy + ity$$
$$= (x + iy)(t + iu) \quad \checkmark$$

Although the transformed formula involves more addition and subtraction operations, by having one less multiplication, this may actually speed up a computer program that involves several calculations with complex numbers. The underlying idea of carrying out the computation with three multiplications rather than four multiplications is used in the Karatsuba algorithm (not just for performing calculations with complex numbers; it is useful for computations involving real numbers, too), designed to make programs more efficient. An alternative way to show that two complex numbers can be multiplied using only three real multiplications is to note that:

$$(x + iy)(t + iu)$$
$$= xt - uy$$
$$+ i(x + y)(t + u)$$
$$- ixt - iyu$$

Here, the three multiplications include xt, uy, and $(x + y)(t + u)$.

Square Roots

Exercises 1, 17, 18, 27, 28, 34, 38, 39, 46, 49, 54, 55, 57, 60, 61, 63, 72, 73, 77, 82, 84, 90, 91, 92, 96, 97, 98

Variable in the Denominator

Exercises 2, 5, 11, 12, 15, 19, 21, 22, 23, 26, 28, 31, 32, 35, 37, 41, 51, 53, 54, 58, 59, 61, 63, 67, 75, 81, 90, 91, 93, 98, 99

System of Equations

Exercises 2, 6, 8, 10, 16, 22, 25, 35, 39, 41, 44, 48, 56, 58, 62, 64, 65, 74, 78, 80, 82, 84, 95, 97, 98, 99

Solve a Single Equation

Exercises 1, 5, 18, 20, 26, 32, 43, 53, 61, 70, 73, 77, 86, 89

Simplify an Expression

Exercises 3, 7, 11, 14, 17, 21, 24, 28, 36, 42, 49, 51, 57, 69, 81, 97, 100

Derive/Show Problems

Exercises 9, 12, 13, 15, 19, 21, 23, 24, 29, 30, 31, 33, 37, 40, 45, 46, 49, 50, 52, 54, 55, 57, 59, 60, 63, 66, 68, 72, 75, 78, 79, 81, 83, 85, 88, 90, 91, 92, 96, 97, 98, 100, 101

Exponent Rules

Exercises 3, 24, 36, 44, 49, 55, 57, 86, 87, 89

Fractional Exponents

Exercises 3, 5, 24, 36, 49, 57, 89

Multiply Expressions

Exercises 3, 7, 9, 14, 20, 33, 34, 36, 43, 45, 70, 85, 91, 96, 101

Inequalities

Exercises 15, 23, 29, 47, 67, 68, 72, 93

Reciprocals

Exercises 2, 11, 12, 32, 33, 35, 51, 58, 67, 76, 83, 85, 93, 99

Polynomials

Exercises 7, 9, 14, 21, 28, 30, 33, 34, 45, 57, 81

Factoring

Exercises 7, 8, 30, 42, 49, 52, 55, 57, 64, 75, 77, 82, 83, 86, 90, 91, 100

Distributing

Exercises 2, 3, 21, 36, 55, 62, 69, 72, 79, 88, 90, 98, 99

Completing the Square

Exercises 30, 86, 97

Partial Fractions

Exercise 64

Transformation of Variables

Exercises 12, 30, 49, 96, 101

Minimum/Maximum

Exercises 54, 79

Separation of Variables

Exercise 49

Cubic Equations

Exercises 33, 70, 96

Quadratic Equations

Exercises 2, 18, 26, 39, 43, 48, 53, 54, 56, 58, 65, 70, 71, 73, 76, 80, 82, 84, 89, 92, 97

The "foil" Method

Exercises 1, 3, 20, 21, 46, 51, 53, 60, 63, 65,
70, 71, 77, 79, 80, 81, 82, 86, 90, 97, 98, 101

Cross Multiply

Exercises 17, 22, 26, 41, 53, 59, 67, 72, 88, 90, 93

Rationalize the Denominator

Exercises 8, 28, 30, 34, 38, 76, 79, 97

Conjugate Expressions

Exercises 8, 28

Common Denominators

Exercises 2, 11, 12, 17, 21, 28, 31, 40, 47, 64, 72, 80, 83, 90

Ratios

Exercises 8, 39, 86, 87, 88

Fractional Coefficients

Exercises 74, 80

Decimals

Exercises 10, 95

Absolute Values

Exercises 63, 95

Complex/Imaginary Numbers

Exercise 101

Fibonacci Sequence

Exercise 11

Straight Lines

Exercises 4, 27, 38, 71, 82, 94

Coordinate Graphs

Exercises 4, 27, 38, 71, 76, 94

Parabola

Exercises 63, 71

Ellipse

Exercises 46, 76

Circles

Exercises 82, 94

Hyperbola

Exercise 76

Cube

Exercise 87

Sphere

Exercise 87

Tangent

Exercise 94

Heron's Formula

Exercise 100

Euler's Formula

Exercise 29

Pythagorean Theorem

Exercise 88

Percent Error

Exercise 95

Applications in Geometry

Exercises 29, 50, 68, 87, 88, 100

Applications in Trigonometry

Exercise 97

Applications in Physics

Exercises 15, 23, 37, 40, 52, 54, 72, 75, 98

Applications in Thermodynamics

Exercises 13, 36, 66, 90-91

Applications in Chemistry

Exercise 13

Applications in Computer Science

Exercise 101

WAS THIS BOOK HELPFUL?

Much effort and thought were put into this book, such as:
- Including a variety of algebra problems that apply a variety of practical skills.
- Modeling some problems after applications in science or higher-level math.
- Providing full step-by-step solutions to all of the problems. Many solutions also include explanations.
- Making separate sections for answers, hints, and full solutions.

If you appreciate the effort that went into making this book possible, there is a simple way that you could show it:

Please take a moment to post an honest review.

For example, you can review this book at Amazon.com or Goodreads.com.

Even a short review can be helpful and will be much appreciated. If you are not sure what to write, following are a few ideas, though it is best to describe what is important to you.
- Was it helpful to have separate sections for answers, hints, and full solutions?
- Were you able to understand the solutions at the back of the book?
- If you were stuck on a problem, was it helpful to read the hints?
- Did this book offer good practice for you? Did the solutions to most of the problems seem to be "involved," as suggested by the title of the book?
- Would you recommend this book to others? If so, why?

Do you believe that you found a mistake? Please email the author, Chris McMullen, at greekphysics@yahoo.com to ask about it. One of two things will happen:
- You might discover that it wasn't a mistake after all and learn why.
- You might be right, in which case the author will be grateful and future readers will benefit from the correction. Everyone is human.

ABOUT THE AUTHOR

Dr. Chris McMullen has over 20 years of experience teaching university physics in California, Oklahoma, Pennsylvania, and Louisiana. Dr. McMullen is also an author of math and science workbooks. Whether in the classroom or as a writer, Dr. McMullen loves sharing knowledge and the art of motivating and engaging students.

The author earned his Ph.D. in phenomenological high-energy physics (particle physics) from Oklahoma State University in 2002. Originally from California, Chris McMullen earned his Master's degree from California State University, Northridge, where his thesis was in the field of electron spin resonance.

As a physics teacher, Dr. McMullen observed that many students lack fluency in fundamental math skills. In an effort to help students of all ages and levels master basic math skills, he published a series of math workbooks on arithmetic, fractions, long division, word problems, algebra, geometry, trigonometry, logarithms, and calculus entitled *Improve Your Math Fluency*. Dr. McMullen has also published a variety of science books, including astronomy, chemistry, and physics workbooks.

Author, Chris McMullen, Ph.D.

MATH

This series of math workbooks is geared toward practicing essential math skills:

- Prealgebra
- Algebra
- Geometry
- Trigonometry
- Logarithms and exponentials
- Calculus
- Fractions, decimals, and percentages
- Long division
- Arithmetic
- Word problems
- Roman numerals
- The four-color theorem and basic graph theory

www.improveyourmathfluency.com

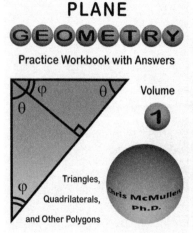

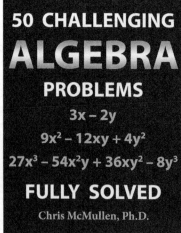

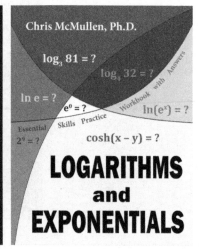

PUZZLES

The author of this book, Chris McMullen, enjoys solving puzzles. His favorite puzzle is Kakuro (kind of like a cross between crossword puzzles and Sudoku). He once taught a three-week summer course on puzzles. If you enjoy mathematical pattern puzzles, you might appreciate:

300+ Mathematical Pattern Puzzles

Number Pattern Recognition & Reasoning

- Pattern recognition
- Visual discrimination
- Analytical skills
- Logic and reasoning
- Analogies
- Mathematics

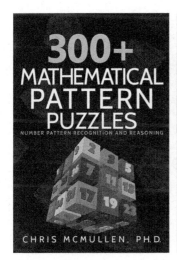

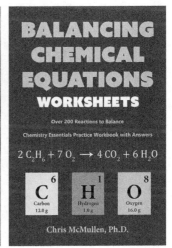

THE FOURTH DIMENSION

Are you curious about a possible fourth dimension of space?

- Explore the world of hypercubes and hyperspheres.
- Imagine living in a two-dimensional world.
- Try to understand the fourth dimension by analogy.
- Several illustrations help to try to visualize a fourth dimension of space.
- Investigate hypercube patterns.
- What would it be like to be a 4D being living in a 4D world?
- Learn about the physics of a possible four-dimensional universe.

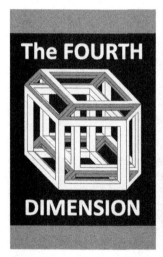

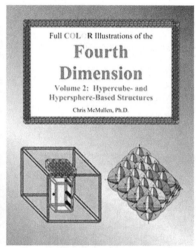

SCIENCE

Dr. McMullen has published a variety of **science** books, including:

- Basic astronomy concepts
- Basic chemistry concepts
- Balancing chemical reactions
- Calculus-based physics textbooks
- Calculus-based physics workbooks
- Calculus-based physics examples
- Trig-based physics workbooks
- Trig-based physics examples
- Creative physics problems
- Modern physics

www.monkeyphysicsblog.wordpress.com

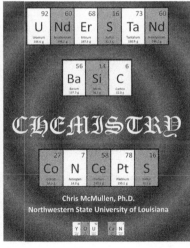

Printed in Great Britain
by Amazon

27836839R00143